Malaria control
in humanitarian emergencies

AN INTER-AGENCY FIELD HANDBOOK

Second Edition

WHO Library Cataloguing-in-Publication Data

Malaria control in humanitarian emergencies: an inter-agency field handbook – 2nd ed.

First edition published with title "Malaria control in complex emergencies: an inter-agency field handbook"

1.Malaria – prevention and control. 2.Emergencies. 3.Malaria – drug therapy. 4.Relief work. 5.Disease outbreaks. 6.Mosquito control. 7.Handbooks. I.World Health Organization.

ISBN 978 92 4 154865 6 (NLM classification: WC 765)

Please consult the WHO Global Malaria Programme web site for the most up-to-date version of all documents (www.who.int/malaria).

Cover photo: © 2008 Paul Jeffrey, Courtesy of Photoshare
Designed by minimum graphics
Printed in Italy

Contents

Tables

Preface

This second edition represents a thorough updating and revision of the first edition. The structure remains similar, but includes an additional chapter on humanitarian coordination. All chapters have been revised to reflect changes in best practices, improvements in technologies, availability of new tools, and changes in WHO recommendations.

The interagency handbook was developed to set out effective malaria control responses in humanitarian emergencies, particularly during the acute phase when reliance on international humanitarian assistance is greatest. It provides policy-makers, planners, and field coordinators with practical advice on designing and implementing measures to reduce malaria morbidity and mortality in both man-made and natural disasters. Such measures must address the needs of all affected population groups and accommodate changing needs as an acute emergency evolves into either recovery or chronic emergency phase. The handbook is organized as follows:

Chapter 1: Introduction introduces complex humanitarian emergencies and malaria control.

Chapter 2: Coordination describes essential coordination, advocacy and resource mobilization.

Chapter 3: Assessment and operational planning describes how to assess malaria burden in an emergency, identify those most at risk, and use the information collected to design an effective response.

Chapter 4: Surveillance discusses establishment of disease surveillance systems to monitor the malaria situation.

Chapter 5: Outbreaks describes how to prepare for and respond to a sudden increase in malaria cases.

Chapter 6: Case management describes methods of diagnosis, treatment, and patient care in humanitarian emergencies.

Chapter 7: Prevention describes approaches and tools for vector control and personal malaria protection during emergencies.

Chapter 8: Community participation discusses how to mobilize affected communities to improve malaria control interventions.

Chapter 9: Operational research and associated routine monitoring discusses conducting research to improve the effectiveness of prevention and treatment in humanitarian emergencies.

A glossary is provided at the beginning of the handbook. Suggestions for further reading are included at the end of several chapters.

Ideal, or *gold standard*, approaches to malaria control are not always feasible in humanitarian emergencies. Interventions must be adapted to the realities of each emergency. Using this handbook should help humanitarian workers implement effective and concerted responses to malaria problems.

As new information becomes available, updates to this handbook will be published. Comments and suggestions are welcome and should be sent to WHO/Global Malaria Programme (infogmp@who.int).

Contributors

This updated second version of the original 2005 interagency handbook is an initiative of the WHO Global Malaria Programme (GMP). In 2011, WHO/GMP brought together the following agencies, involved in malaria control in humanitarian emergencies, to revise and update this handbook:

Centers for Disease Control and Prevention (CDC), USA
HealthNet-TPO, Netherlands
London School of Hygiene & Tropical Medicine (LSHTM), United Kingdom
Malaria Consortium, United Kingdom
Médecins Sans Frontières (MSF), Belgium
MENTOR Initiative, United Kingdom
National Malaria Control Programme, Ministry of Health, Democratic Republic of the Congo (DRC)
Office of the United Nations High Commissioner for Refugees (UNHCR), Switzerland
United Nations Children's Fund (UNICEF):
  Headquarters, USA
  Regional Office, Eastern and Southern Africa (ESARO), Kenya
World Health Organization (WHO):
  Headquarters, Switzerland
  Regional Office, Africa (AFRO)
  Country Office, South Sudan

A writing committee of malaria experts coordinated by Jose Nkuni, Natasha Howard and Abraham Mnzava produced the draft of this second version. Contributors are listed alphabetically per chapter:

Chapters 1–3
Michelle Gayer (WHO/DCE)
Anne Golaz (UNICEF)
Natasha Howard (LSHTM)
Marian Schilperoord (UNHCR)

Chapters 4–5
Martin de Smet (MSF-Belgium)
Georges Ki-Zerbo (WHO/AFRO)
Toby Leslie (Healthnet-TPO)
Michael Lynch (WHO/GMP and CDC)
Aafje Rietveld (WHO/GMP)
David Townes (CDC/IERH)

Chapter 6
Jean Angbalu (NMCP DRC)
Andrea Bosman (WHO/GMP)
Michelle Chang (CDC)
Jane Cunningham (WHO/GMP)
Asis Das (UNHCR)
Luz Maria de Regil (WHO/NHD)
Prudence Hamade (Malaria Consortium)
Natasha Howard (LSHTM)
Jeylani Mohammoud (WHO South Sudan)
Peter Ehizibue Olumese (WHO/GMP)
Marian Warsame (WHO/GMP)

Chapter 7
Richard Allan (Mentor Initiative)
Natasha Howard (LSHTM)
Jo Lines (LSHTM)
Michael Macdonald (WHO/GMP)
Peter Maes (MSF-Belgium)
Abraham Mnzava (WHO/GMP)
Jose Nkuni RBM/CARN (former WHO/GMP)
Melanie Renshaw (ALMA)
Mark Rowland (LSHTM)

Chapter 8
Heather Popowitz (UNICEF)
Holly Williams (CDC)

Chapter 9
Richard Allan (Mentor Initiative)
Natasha Howard (LSHTM)
Michael Lynch (WHO/GMP)
Michael Macdonald (WHO/GMP)
Peter Maes (MSF-Belgium)
Melanie Renshaw (ALMA)
Mark Rowland (LSHTM)

The handbook was edited by Natasha Howard (LSHTM), with additional technical editing provided by Alison Clements-Hunt. Erin Shutes coordinated with agencies for consensus and Robert Newman provided the final technical review for the WHO Global Malaria Programme.

Acknowledgements

The editor and writing committee gratefully acknowledge all the colleagues who reviewed chapters before publication, particularly the specialists who contributed to further development of the text. Thanks are due to the United Nations agencies, non-governmental organizations, and operational field partners of GMP for their support.

Finally, we thank the Office of U.S. Foreign Disaster Assistance of the United States Agency for International Development (USAID/OFDA) and WHO for generous financial support for this publication.

Abbreviations

ALNAP	Active Learning Network for Accountability and Performance
AFPMB	Armed Forces Pest Management Board
ANC	antenatal care
ACT	artemisinin-based combination therapy
ANVR	African Network on Vector Resistance
BCC	behaviour change communications
CDC	Centers for Disease Control and Prevention (United States)
CERF	Central Emergency Response Fund
CHAP	Common Humanitarian Action Plan
CHW	community health worker
CMR	crude mortality rate
CSF	cerebrospinal fluid
DFID	Department for International Development (United Kingdom)
DRC	Democratic Republic of the Congo
EPI	Expanded programme on immunization
ERC	Emergency Response Coordinator
EVF	erythrocyte volume fraction (haematocrit)
GPIRM	Global Plan for Insecticide Resistance Management in malaria vectors
GPS	global positioning system
HC	Humanitarian Coordinator
HPLC	high-performance liquid chromatography
IASC	United Nations Inter-Agency Standing Committee
IDP	internally displaced person
IEC	information, education and communication
IEHK	interagency emergency health kits
IFRC	International Federation of Red Cross and Red Crescent Societies
IM	intramuscular
IMCI	Integrated Management of Childhood Illness
IPT	intermittent preventive treatment
IPTi	intermittent preventive treatment in infants

IPTp	intermittent preventive treatment in pregnancy
IRC	International Rescue Committee
IRS	indoor residual spraying
ITM	insecticide-treated material
ITN	insecticide-treated net
ITPS	insecticide treated plastic sheeting
IV	intravenous
LP	lumbar puncture
LLINs	long lasting insecticidal nets
MOH	ministry of health
MPS	making pregnancy safer
MSF	Médecins Sans Frontières
NG	nasogastric
NGO	non-governmental organization
NMCP	National Malaria Control Programme
OFDA	Office of U.S. Foreign Disaster Assistance
ORS	oral rehydration solution
PLHIV	people living with HIV
PMI	President's Malaria Initiative
PSM	procurement and supply chain management
PVO	private voluntary organization
RC	resident coordinator
RBM	Roll Back Malaria
RDT	rapid diagnostic test
SMC	seasonal malaria chemoprevention
SP	sulfadoxine–pyrimethamine
TFC	therapeutic feeding centre
UNDAF	United Nations Development Assistance Framework
UNDHA	United Nations Department for Humanitarian Affairs
UNDMT	United Nations Disaster Management Teams
UNDP	United Nations Development Programme
UNICEF	United Nations Children's Fund
UNISDR	United Nations Office for Disaster Risk Reduction
UNHCR	Office of the United Nations High Commissioner for Refugees
UPS	untreated plastic sheeting
USAID	United States Agency for International Development
WASH	water, sanitation and hygiene
WFP	World Food Programme
WHO	World Health Organization

Glossary

The definitions given below apply to the terms as used in this handbook. They may have different meanings in other contexts.

A

acute emergency phase Begins immediately after the impact of the disaster and may last for up to 3 months. Characterized by initial chaos and a high crude mortality rate (CMR) and ends when daily CMR drops below 1/10 000 people.

adherence (compliance) Health-related behaviour that abides by the recommendations of a doctor or other health care provider or of an investigator in a research project.

agranulocytosis Severe deficiency of certain white blood cells as a result of damage to the bone marrow by toxic drugs or chemicals.

anaemia A reduction in the quantity of the oxygen-carrying pigment haemoglobin in the blood. The main symptoms are tiredness, breathlessness on exertion, pallor and poor resistance to infection.

Anopheles A genus of widely distributed mosquitoes, occurring in tropical and temperate regions and containing some 400 species. Malaria parasites (*Plasmodium*) are transmitted to humans through the bite of female *Anopheles* mosquitoes.

***Anopheles*, infected** Female *Anopheles* with oocysts of malaria parasites on the midgut wall (with or without sporozoites in the salivary glands).

***Anopheles*, infective** Female *Anopheles* with sporozoites in the salivary glands (with or without oocysts in the midgut).

antipyretic A drug that reduces fever by lowering the body temperature.

anthropophilic Descriptive of mosquitoes that show a preference for feeding on humans even when non-human hosts are available. A relative term requiring quantification to indicate the extent of the preference.

asymptomatic Not showing any symptoms of disease, whether disease is present or not. (See parasitaemia for "asymptomatic malaria".)

attack rate The cumulative incidence of infection in a group observed over a period during an epidemic.

auscultation Listening, usually with the aid of a stethoscope, to the sounds produced by the movement of gas/air or fluid within the body, to diagnose abnormalities, for example in the lungs.

B

bacteraemia Presence of bacteria in the blood.

bias Any trend in the collection, analysis, interpretation, publication or review of data that can lead to conclusions that are systematically different from the truth.

blood meal Ingestion by a female mosquito of blood obtained from a vertebrate host; also, the ingested blood.

bradycardia Slowing of the heart rate to less than 50 beats per minute.

breeding site (place) Site where eggs, larvae or pupae of mosquitoes are found; larval habitat.

C

case A person who has the particular disease, health disorder or condition that meets the case definition.

case definition A set of diagnostic criteria that must be fulfilled for an individual to be regarded as a "case" of a particular disease for surveillance and outbreak investigation purposes. Case definitions can be based on clinical criteria, laboratory criteria or a combination of the two.

case-fatality rate The proportion of cases of a specified condition that are fatal within a specified time (usually expressed as a percentage).

census Enumeration of a population. A census usually records the identities of all persons in every place of residence, with age or date of birth, sex, occupation, national origin, language, marital status, income, and relationship to head of household, in addition to information on the dwelling place.

chemoprophylaxis Administration of a chemical to prevent either the development of an infection or the progression of an infection to active manifest disease.

cluster sampling A sampling method in which each selected unit is a group of people (e.g. all persons in a city block, a family) rather than an individual.

community health worker (CHW) A member of the community who is integrated into primary health care programmes after a short training on health-related issues, and who acts as intermediary between the community and the health services. CHWs may be paid staff or volunteers (CHVs).

compliance See **adherence.**

confidence interval A computed interval with a given probability (e.g. 95%) that the true value of a variable (e.g. a mean, proportion or rate) is contained within that interval.

conjunctival Relating to the conjunctiva, the thin mucosa covering the inside of the eyelids and the sclera (the white part of the eye).

contact (of an infection) A person or animal that has been in such association with an infected person or animal or a contaminated environment as to have had opportunity to acquire the infection.

coverage A measure of the extent to which services cover the potential need for these services in a community. It is expressed as a proportion in which the numerator is the number of services rendered and the denominator is the number of instances in which the service should have been rendered.

cross-sectional study (disease frequency survey; prevalence study) A study that examines the relationship between diseases (or other health-related characteristics) and other variables of interest as they exist in a defined population at one particular time. The presence or absence of disease and the presence or absence (or, if they are quantitative, the level) of the other variables are determined in each member of the study population or in a representative sample at one particular time. The relationship between a variable and the disease can be examined in terms both of the prevalence of disease in different population subgroups defined according to the presence or absence (or level) of the variables and of the presence or absence (or level) of the variables in the diseased versus the non-diseased. A cross-sectional study usually records disease prevalence rather than incidence and cannot necessarily determine cause-and-effect relationships.

D

demography The study of populations, especially with reference to size and density, fertility, mortality, growth, age distribution, migration and vital statistics, and the interaction of all these with social and economic conditions.

district hospital A hospital with the capacity to manage first-referral cases but with medical services usually limited to emergency obstetrical and surgical care and follow-up, and inpatient and rehabilitative care. In principle, facilities include laboratory, blood bank, and X-ray services.

E

effectiveness A measure of the extent to which a specific intervention, procedure, regimen or service, when deployed in the field in routine circumstances, does what it is intended to do for a specified population; a measure of the extent to which a health care intervention fulfils its objectives.

efficacy The extent to which a specific intervention, procedure, regimen or service produces a beneficial result under ideal conditions; the benefit or utility to the individual or population of the service, treatment regimen or intervention. Ideally, determination of efficacy is based on the results of a randomized controlled trial.

efficiency The effects or end results achieved in relation to the effort expended in terms of money, resources and time; the extent to which the resources used to provide a specific intervention, procedure, regimen or service of known efficacy and effectiveness are minimized; a measure of the economy (or cost in resources) with which a procedure of known efficacy and effectiveness is carried out; the process of making the best use of scarce resources.

endemic Description applied to malaria when there is a constant measurable incidence both of cases and of natural transmission in an area over a succession of years.

endophagy Tendency of mosquitoes to feed indoors.

endophily Tendency of mosquitoes to rest indoors.

epidemic Description applied to malaria when the incidence of cases (other than seasonal increases) in an area rises rapidly and markedly above its usual level or when the infection occurs in an area where it was not previously present.

epidemic curve A graphic plotting of the distribution of cases by time of onset.

epidemiology The study of the distribution and determinants of health-related conditions or events in specified populations, and the application of this study to the control of health problems.

essential drugs Therapeutic substances that are indispensable for rational care of the majority of diseases in a given population.

evaluation A process that attempts to determine as systematically and objectively as possible the relevance, effectiveness and impact of activities in relation to their objectives.

exophagy Tendency of mosquitoes to feed outdoors.

exophily Tendency of mosquitoes to rest outdoors.

F

focus group Small convenience sample of people brought together to discuss a topic or issue with the aim of ascertaining the range and intensity of their views, rather than arriving at a consensus. Focus groups are used, for example, to appraise perceptions of health problems, assess the acceptability of a field study or refine the questions to be used in a field study.

G

gamete Mature sexual form, male or female. In malaria parasites, the female gametes (macrogametes) and the male gametes (microgametes) normally develop in the mosquito.

gametocytes Parent cell of a gamete. In malaria parasites, the female and male gametocytes develop in the red blood cell. Very young gametocytes cannot usually be distinguished from trophozoites.

H

haemoglobinuria The presence of haemoglobin in the urine

haemolysis The destruction of red blood cells.

health care system The organization of health care services within a designated geographical area (e.g. country, province, district, camp).

hepatic Relating to the liver.

host A person or other living animal, including birds and arthropods, harbouring or providing subsistence to a parasite. In an epidemiological context, the host may be the population or group; biological, social and behavioural characteristics of this group that are relevant to health are called "host factors".

host, definitive In parasitology, the host in which sexual maturation occurs. In malaria, the definitive host is the mosquito (invertebrate host).

host, intermediate In parasitology, the host in which asexual forms of the parasite develop. In malaria, the intermediate host is a human or other mammal or bird (vertebrate host).

household One or more persons who occupy a dwelling; may or may not be a family. The term is also used to describe the dwelling unit in which the persons live.

hypersensitivity Responding exaggeratedly to the presence of a particular antigen.

I

immunity, acquired Resistance acquired by a host as a result of previous exposure to a natural pathogen or substance that is foreign to the host.

incidence The number of instances of illness commencing, or of persons falling ill, during a given period in a specified population.

incidence rate The rate at which new events occur in a population. The numerator is the number of new events that occur in a defined period and the denominator is the population at risk of experiencing these events during this period, sometimes expressed as person-time.

indicator A measure that shows whether a standard has been reached. It is used to assess and communicate the results of programmes as well as the process or methods used. Indicators can be qualitative or quantitative.

infection, mixed Malaria infection with more than one species of *Plasmodium*.

informed consent Voluntary consent given by a subject (i.e. person or a responsible proxy such as a parent) for participation in a study, immunization programme, treatment regimen, etc., after being informed of the purpose, methods, procedures, benefits and risks, and, when relevant, the degree of uncertainty about outcomes. The essential criteria of informed consent are that the subject has both knowledge and comprehension, that consent is freely given without duress or undue influence, and that the right of withdrawal at any time is clearly communicated to the subject.

inpatient facility A health care institution that provides lodging, nursing and continuous medical care for patients within a facility and with an organized professional staff.

K

knowledge, attitudes, practice (KAP) survey A formal survey, using face-to-face interviews, in which people are asked standardized pre-tested questions dealing with their knowledge, attitudes and practice concerning a given health or health-related problem.

knockdown Rapid immobilization of an insect by an insecticide, without necessarily causing early death.

L

larva The pre-adult or immature stage hatching from a mosquito egg.

loading dose An initial higher dose of a drug, given with the objective of rapidly providing an effective drug concentration.

logistics The procurement, maintenance and transport of material, personnel and facilities; management of the details of an undertaking.

M

merozoite A stage in the life cycle of the malaria parasite. Product of segmentation of a tissue schizont, or of an erythrocytic schizont before entering a new host cell. Merozoites are found either separated from or contained in the original schizont.

mesoendemicity A pattern of malaria transmission typically found among small rural communities in subtropical zones, with varying intensity of transmission depending on local circumstances.

mobile clinic See **outreach clinic**.

monitoring Episodic measurement of the effect of an intervention on the health status of a population or environment; not to be confused with surveillance, although surveillance techniques may be used in monitoring. The process of collecting and analysing information about the implementation of a programme for the purpose of identifying problems, such as non-compliance, and taking corrective action. In management: the episodic review of the implementation of an activity, seeking to ensure that inputs, deliveries, work schedules, targeted outputs and other required actions are proceeding according to plan.

morbidity Any departure, subjective or objective, from a state of physiological or psychological well-being. *Sickness*, *illness*, and *morbid condition* are similarly defined and synonymous. Morbidity can be measured as: (a) persons who are ill; (b) the illnesses experienced by these persons; and (c) the duration of these illnesses.

mortality rate An estimate of the proportion of a population dying during a specified period. The numerator is the number of persons dying during the period and the denominator is the total number of the population, usually estimated as the mid-year population.

myalgia Pain in the muscles.

N

needs assessment A systematic procedure for determining the nature and extent of problems that directly or indirectly affect the health of a specified population. Needs assessment makes use of epidemiological, socio-demographic and qualitative methods to describe health problems and their environmental, social, economic and behavioural determinants. The aim is to identify unmet health care needs and make recommendations about ways to address these needs.

O

oedema Swelling caused by an excess of fluid in the tissues.

oliguria The production of an abnormally small amount of urine over a period of time.

outpatient clinic A facility for diagnosis, treatment and care of ambulatory patients.

outreach (clinic, team) The extending of services or assistance beyond fixed facilities.

P

pallor Paleness.

palmar Relating to the palms of the hands.

parasitaemia Condition in which malaria parasites are present in the blood. If this condition in the human subject is not accompanied by fever or other symptoms of malaria except for a possible enlargement of the spleen, it is known as asymptomatic parasitaemia, and the person exhibiting the condition is known as an asymptomatic parasite carrier.

parenteral Relates to drug administration by any route other than oral, e.g. by injection.

paroxysms Cyclic manifestation of acute illness in malaria characterized by a rise in temperature with accompanying symptoms; usually caused by invasion of the blood by a brood of erythrocytic parasites.

Plasmodium Genus of parasites causing human malaria (*Plasmodium falciparum*, *P. vivax*, *P. ovale*, *P. malariae*).

population structure Composition of the population, usually by age and sex.

preparedness Readiness to prevent, mitigate, respond to and cope with the effects of a disaster or epidemic.

prevalence The number of instances of a given disease or other condition in a given population at a designated time, with no distinction between new and old cases. Prevalence may be recorded at a specific moment (point prevalence) or over a given period of time (period prevalence).

prevalence rate The total number of all individuals who have a disease or condition at a particular time (or during a particular period) divided by the population at risk of having the attribute or disease at that point in time or midway through that period.

prevention Actions aimed at eradicating, eliminating or minimizing the impact of disease and disability or, if none of these is feasible, retarding the progress of disease and disability. The concept of prevention is best defined in the context of levels, called primary, secondary, and tertiary prevention. In epidemiological terms, primary prevention aims to reduce the incidence of disease; secondary prevention aims to reduce the prevalence of disease by shortening its duration; and tertiary prevention aims to reduce the number and/or impact of complications.

proportion See under **ratio**.

pulmonary Relating to, associated with or affecting the lungs.

R

random sample A sample that is arrived at by selecting sample units in such a way that each possible unit has a fixed and determinate probability of selection. Random allocation follows a predetermined plan that is usually devised with the aid of a table of random numbers. Random sampling should not be confused with haphazard assignment.

rate A measure of the frequency of occurrence of a phenomenon. In epidemiology, demography and vital statistics, a rate is an expression of the frequency with which an event occurs in a defined population in a specified period of time. The use of rates rather than raw numbers is essential for comparison of experience between populations at different times, different places, or among different groups of persons. The components of a rate are the numerator, the denominator, the specified time in which events occur, and usually a multiplier, a power of 10, that converts the rate from an awkward fraction or decimal to a whole number. E.g. number of events in specified period divided by average population during the period

ratio The value obtained by dividing one quantity by another; a general term of which rate, proportion, percentage, etc. are sub-sets. The important difference between a proportion and a ratio is that the numerator of a proportion is included in the population defined by the denominator, whereas this is not necessarily so for a ratio. A ratio is an expression of the relationship between a numerator and a denominator where the two are usually separate and distinct quantities, neither being included in the other.

recrudescence Repeated manifestation of an infection after a period of latency following the primary attack. It is used particularly in the context of treatment failure of *Plasmodium falciparum*, and is often the result of non-adherence to the treatment regimen, especially with short-acting drugs such as quinine and the artemisinins, and can signify antimalarial drug resistance.

relapse Renewed manifestation (of clinical symptoms and/or parasitaemia) of malaria infection separated from previous manifestations of the same infection by an interval greater than that related to the normal periodicity of the paroxysms. The term is used mainly for renewed manifestation due to survival of exo-erythrocytic forms of *Plasmodium vivax* or *P. ovale*.

reservoir (of infection) Any person, animal, arthropod, plant, soil or substance, or a combination of these, in which an infectious agent normally lives and multiplies, on which it depends primarily for survival, and where it reproduces itself in such a manner that it can be transmitted to a susceptible host. The natural habitat of the infectious agent.

resistance Ability of a parasite strain to multiply or to survive in the presence of concentrations of a drug that normally destroy parasites of the same species or prevent their multiplication. Ability in a population of insects to tolerate doses of an insecticide that would prove lethal to the majority of individuals in a normal population of the same species; developed as a result of selection pressure by the insecticide.

retention (in ITN programme) An indicator used to establish whether nets remain with the individuals to whom they were originally distributed. Retention alone is not an indicator of the correct use of the nets.

risk The probability that an event will occur, e.g. that an individual will become ill or die within a stated period of time or by a certain age. Also a non-technical term encompassing a variety of measures of the probability of a (generally) unfavourable outcome.

S

sample A selected sub-set of a population. A sample may be random or non-random and may be representative or non-representative.

sampling The process of selecting a number of subjects from all subjects in a particular group, or "universe". Conclusions based on sample results may be attributed only to the population from which the sample was taken. Any extrapolation to a larger or different population is a judgement or a guess and is not part of statistical inference.

schizont Intracellular asexual form of the malaria parasite, developing either in tissue or in blood cells.

sensitivity (of a screening test) The proportion of truly diseased persons in the screened population who are identified as diseased by the screening test. Sensitivity is a measure of the probability of correctly diagnosing a case, or the probability that any given case will be identified by the test. See also specificity.

species Group of organisms capable of exchanging genetic material with one another and incapable, by reason of their genetic constitution, of exchanging such material with any other group of organisms.

specificity (of a screening test) The proportion of truly non-diseased persons who are so identified by the screening test. Specificity is a measure of the probability of correctly identifying a non-diseased person with a screening test.

spleen rate A term used in malaria epidemiology to define the frequency of enlarged spleens detected on survey of a population in which malaria is prevalent.

splenomegaly Enlargement of the spleen.

sporozoite The infective form of the malaria parasite occurring in a mature oocyst before its rupture and in the salivary glands of the mosquito.

surveillance The process of systematic collection, orderly consolidation and evaluation of pertinent data with prompt dissemination of the results to those who need to know, particularly those who are in a position to take action.

survey An investigation in which information is systematically collected; usually carried out in a sample of a defined population group, within a defined time period. Unlike surveillance a survey is not continuous; however, if repeated regularly, surveys can form the basis of a surveillance system.

systematic sampling The procedure of selecting according to some simple, systematic rule, such as all persons whose names begin with specified alphabetical letters, who are born on certain dates, or who are located at specified points on a master list. A systematic sample may lead to errors that invalidate generalizations. For example, people's names more often begin with certain letters of the alphabet than with other letters. A systematic alphabetical sample is therefore likely to be biased.

T

target population The group of people for whom an intervention is intended.

transmission Any mechanism by which an infectious agent is spread from a source or reservoir to another person.

transmission, perennial Transmission occurring throughout the year without great variation of intensity.

transmission, seasonal Natural transmission that occurs only during some months and is totally interrupted during other months.

trend A long-term movement in an ordered series, e.g. a time series. An essential feature is movement consistently in the same direction over the long term.

triage The process of selecting for care or treatment those of highest priority or, when resources are limited, those thought most likely to benefit.

trophozoite Strictly, any asexual and growing parasite with an undivided nucleus. In malaria terminology, generally used to indicate intracellular erythrocytic forms in their early stages of development. Trophozoites may be in either a ring stage or an early amoeboid or solid stage, but always have the nucleus still undivided.

V

vector Any insect or living carrier that transports an infectious agent from an infected individual or his/her waste to a susceptible individual or his/her food or immediate surroundings. The organism may or may not pass through a developmental cycle within the vector.

vector control Measures of any kind directed against a vector of disease and intended to limit its ability to transmit the disease.

verbal autopsy A procedure for gathering systematic information that enables the cause of death to be determined in situations where the deceased has not been medically attended. It is based on the assumption that most common and important causes of death have distinct symptom complexes that can be recognized, remembered and reported by lay respondents.

vulnerability Defencelessness, insecurity, exposure to risks, shock, and stress, and having difficulty coping with them. The potential that when something destructive happens or goes wrong, people will not be able to handle the consequences by themselves and the ability to sustain life is endangered.

Z

zoophilic Descriptive of mosquitoes showing a relative preference for non-human blood even when human hosts are readily available.

CHAPTER 1

Introduction

This chapter:
- defines humanitarian emergencies, including different types and phases
- describes malaria, including parasites and transmission
- identifies the importance of malaria in humanitarian emergencies and main populations at risk

Humanitarian emergencies

Definitions

A **humanitarian emergency**, as used in this handbook, is equivalent to a major disaster – i.e. a calamitous situation in which the functioning of a community or society is severely disrupted, causing human suffering and material loss that exceeds the affected population's ability to cope using its own resources (UNISDR 2004). This includes natural disasters (e.g. earthquakes) and man-made disasters (e.g. industrial accidents, violent armed conflicts). A **complex emergency**, as described by the United Nations Inter-Agency Standing Committee (IASC) and the Active Learning Network for Accountability and Performance (ALNAP), is a humanitarian emergency with complex social, political and economic origins, breakdown of governmental authority and structures, and often human rights abuses and armed conflict (IASC 1994 and ALNAP 2003).

This second edition will address malaria control in humanitarian emergencies generally.

Types of humanitarian emergencies

Both **natural disasters** and **man-made disasters** can lead to population displacements, food scarcity, and health systems disruption, causing excess mortality and morbidity in affected populations. As the differences between these two types of emergencies seldom affect basic malaria control, this handbook will not generally distinguish between them.

Rapid-onset emergencies describe a suddenly deteriorating situation, such as the Haiti earthquake of 2011. **Slow-onset emergencies** involve a gradually deteriorating situation, such as the Pakistan floods of 2010. **Chronic** or **protracted emergencies** involve consistently high levels of humanitarian need, and often lower levels of aid, as the root causes of the emergency are not being resolved (e.g. Somalia, Afghanistan). These distinctions can sometimes affect malaria control and will be mentioned where likely to be relevant.

An increasing number of humanitarian emergencies involve **open settings**, in which internally displaced populations (IDPs) or refugees settle in host communities and urban settings. Humanitarian assistance is more complicated in open than in traditional **closed settings** (i.e. refugee camps), usually requiring at least an expansion of assistance into host communities to address host community health needs as well as those of refugees or IDPs. These different types of settings can affect malaria control options and will be addressed in this handbook whenever they are likely to be relevant.

Phases of humanitarian emergencies

Humanitarian emergencies often evolve rapidly and unpredictably. Dividing the emergencies into phases can help guide how humanitarian aid should be implemented. These phases generally consist of acute, post-acute, early recovery, and reconstruction phases. Table 1.1 summarizes different phases of a humanitarian emergency.

Table 1.1 **Phases of humanitarian emergencies**

<table>
<tr><th></th><th>Sudden-onset crisis</th><th>Slow-onset crisis</th><th>Chronic crisis</th></tr>
<tr><td>Phase 1</td><td>First 24–72 hours</td><td>First 1–2 weeks</td><td rowspan="2">Ongoing (often low-level conflict with ad-hoc flares)</td></tr>
<tr><td>Phase 2</td><td>First 4–10 days</td><td>First month</td></tr>
<tr><td>Phase 3</td><td>4 to 6 weeks for a disaster
Up to 3 months for conflict</td><td>2 –3 months</td><td rowspan="2">Indefinite (low-level continuation response)</td></tr>
<tr><td>Phase 4</td><td colspan="2">Continuation response and progressive recovery</td></tr>
</table>

Source: adapted from WHO, 2009.

In a humanitarian emergency, key characteristics of the acute phase (i.e. phases 1–2 in the table above) include:

- elevated mortality rates (e.g. crude mortality rate over 1/10 000 population per day, and under-5 mortality rate over 2/10 000 per day) or a doubling of the baseline rate;

- poor access to effective health care for the affected population (e.g. health infrastructure may be overwhelmed, inadequate or non-existent);
- an appropriate response that is beyond local or national capacity;
- the possible breakdown of normal coordination mechanisms.

Malaria

Overview

Malaria is a common and life-threatening disease in many tropical and subtropical areas and it is currently endemic in 99 countries. In 2010, there were an estimated 219 million malaria episodes (uncertainty range 154–289 million), of which approximately 81% were in Africa, and an estimated 660 000 malaria deaths (uncertainty range 490 000 to 836 000), of which 91% were in Africa. Approximately 86% of malaria deaths globally are among children under five years old and an estimated 10 000 pregnant women and 200 000 newborn babies die annually due to malaria during pregnancy. Estimated malaria incidence has been reduced by 17% and malaria-specific mortality rates by 26% globally since 2000. These rates of decline are lower than the 50% target reductions agreed internationally for 2010, but nonetheless represent a major achievement (WHO, 2012).

Parasites

Malaria is caused by the protozoan parasite *Plasmodium*. Human malaria is caused by four different species of *Plasmodium*: *P. falciparum*, *P. malariae*, *P. ovale* and *P. vivax*. Humans occasionally become infected with *Plasmodium* species that normally infect animals, such as *P. knowlesi*. As previously mentioned, over 80% of malaria cases and 90% of malaria deaths occur in tropical sub-Saharan Africa where *P. falciparum* predominates. The current distribution of malaria in the world is shown in Figure 1.1.

Malaria is an acute febrile illness with an incubation period of seven days or longer. The most severe form is caused by *P. falciparum*; variable clinical features include fever, chills, headache, muscular aching and weakness, vomiting, cough, diarrhoea and abdominal pain. Other symptoms related to organ failure may supervene, such as acute renal failure, pulmonary oedema, generalized convulsions, and circulatory collapse, followed by coma and death. The initial symptoms, which can be mild, may not be easy to recognize as being due to malaria. Children, pregnant women, and those with compromised immune systems, such as people living with HIV/AIDS, are especially vulnerable to malaria. Malaria, particularly *P. falciparum*, in non-immune pregnant women increases the risk of maternal death, miscarriage, stillbirth and neonatal death.

Figure 1.1 **Malaria countries or areas at risk of transmission, 2011**

Source: © WHO 2012.

The forms of human malaria caused by *P. vivax*, *P. malariae* and *P. ovale* cause significant morbidity but not the high mortality rates seen with falciparum malaria. Vivax malaria is frequently associated with high fevers, anaemia and splenomegaly. Cases of severe *P. vivax* malaria have been reported among populations living in (sub)tropical countries or areas at risk.

Plasmodium vivax and *P. ovale* can remain dormant in the liver. Relapses caused by these persistent liver forms ("hypnozoites") may appear months, and rarely several years, after exposure. Latent blood infection with *P. malariae* may be present for many years, but it is very rarely life threatening. Human malaria due to *P. knowlesi*, which may be severe, is mainly a public health problem among populations living or working in forested areas in South-East Asia.

Human malaria species are not evenly distributed across the malaria-affected areas of the world; and their relative importance varies between and within different regions. The risk of contracting malaria is therefore highly variable from country to country and even between areas within a country. It is important to know which species are present in any particular area and their relative proportion, because this will affect the appropriate diagnostic modalities and treatment. In Afghanistan, for example, 85% of malaria is caused by *P. vivax* and up to 15% by *P. falciparum*; in DRC, 95% of malaria is due to *P. falciparum*, with *P. malariae* and *P. ovale* accounting for the remaining 5%; in Timor-Leste, 60% is caused by *P. falciparum* and 40% by *P. vivax*.

A global distribution map is available at: http://gamapserver.who.int/mapLibrary/Files/Maps/Global_Malaria_ITHRiskMap.JPG.

A description of malaria risk by country is available at: World Malaria Reports (www.who.int/malaria) and http://www.who.int/entity/ith/chapters/ith2012en_countrylist.pdf.

Vectors

Malaria parasites are transmitted by female Anopheles mosquitoes, which bite mainly between sunset and sunrise. There are approximately 400 different species of Anopheles mosquitoes throughout the world, but only about 60 of these are natural malaria vectors and only 30 are vectors of major importance. Each species behaves differently and knowledge of relevant behaviour patterns is important for planning malaria prevention measures (see Chapter 7).

Epidemiology and population risk

Knowledge of malaria epidemiology in areas affected by humanitarian emergencies is essential for appropriate prevention and case-management. Endemicity refers to the malaria burden in an area, measured as the prevalence of peripheral blood stage infections in a population. Epidemic malaria signifies a periodic sharp increase in incidence. WHO classifies endemicity as hypoendemic (child parasite rates below 10%), mesoendemic, hyperendemic, or holoendemic (child parasite rates above 75%). Transmission can be categorized as stable, with minimal fluctuations over the years (though seasonal fluctuations may occur), or unstable, in which transmission fluctuates from year to year.

Malaria epidemiology varies by region, determining levels of immunity and risk of severe disease. For example, in rural lowland areas of Africa, malaria transmission may be high and stable, whether perennial or seasonal. People in areas of moderate or high transmission gradually develop partial immunity after childhood. Thus, young children are at greatest risk from severe malaria. Malaria in pregnancy can cause maternal anaemia, spontaneous abortions, stillbirths and low birth-weight (particularly in first and second pregnancies and among HIV-infected women of all gravidities). In such settings, incidence of symptomatic disease generally remains relatively constant throughout the year and people may carry malaria parasites in their blood without showing clinical symptoms of disease. Thus, with a high proportion of the population parasite-positive at any time, the presence of parasites in a patient's blood is not a clear indicator that malaria is causing the presenting illness. However, Plasmodium infection, once identified, should always be treated, regardless of whether or not it is the cause of the acute illness.

Seasonal or epidemic peaks of malaria occur in parts of Asia and the Americas where transmission is low to moderate. Transmission can be high in forested and forest-fringe areas of South-East Asia, while population immunity remains low. Anecdotal data indicates that severe disease occurs most frequently in adults staying overnight in forests (e.g. loggers). In high-altitude and desert-fringe areas and city centres, malaria endemicity is usually low, with less than 10% of the population infected, although there may be more intense transmission in certain African and south Asian cities. In areas of unstable malaria transmission, such as highland areas in sub-Saharan Africa and the semi-arid and desert fringes of countries in the Horn of Africa and the Sahel, populations may have low to no immunity and outbreaks of malaria may affect all age and population groups. In these areas, presence of parasites in a patient's blood indicates that malaria is highly likely to be the cause of the presenting illness.

Environmental and political changes can contribute to the emergence or re-emergence of malaria in areas where it was previously non-existent or well controlled. For example, population displacement from malaria-endemic areas of Afghanistan contributed to the re-emergence of *P. falciparum* malaria in Tajikistan. Climate change may be changing the distribution of malaria vectors, and therefore malaria transmission patterns and malaria burden globally. Warmer winter temperatures and prolonged amplification cycles may allow the establishment of imported mosquito-borne diseases in countries from which they have previously been absent. Increased rains and extreme precipitation events can lead to increased mosquito breeding sites, and therefore increased transmission risk.

Importance of malaria in humanitarian emergencies

Falciparum malaria can be rapidly fatal and is a priority during the acute phase of an emergency. Effective malaria control programmes prevent malaria transmission by promoting personal protection and effective vector control, and providing appropriate case management with early diagnostic testing of suspected malaria and effective treatment for those with confirmed infection. However, malaria control in humanitarian emergencies is often complicated by the breakdown of existing health services and programmes, displacement of health workers and field staff with malaria expertise, movement of non-immune people to endemic areas, and concentrations of people, often already in poor health, in high-risk, high-exposure settings.

Regions whose populations are most affected by humanitarian emergencies are often those with the greatest malaria burden. Consequently, malaria is a significant cause of death and illness in many emergencies.

Vulnerability and constraints

Many factors contribute to increased malaria burden in populations affected by humanitarian emergencies, including:

- the collapse of health services;
- ongoing conflict limiting access to effective treatment;
- resistance of malaria parasites to commonly available drugs and vector resistance to insecticides;
- lack of immunity to malaria due to population displacement from non-malarious or low-transmission areas to/through an area of high transmission (Table 1.2);

— weakened immunity because of multiple infections and malnutrition;
— increased exposure to *Anopheles* mosquitoes due to poor or absent housing;
— environmental deterioration resulting in increased vector breeding.

Women and young children often make up the largest proportion of malaria-affected populations. Additionally, some groups are particularly vulnerable, especially in the acute phase of an emergency, because they are given lower priority in the distribution of limited resources, are marginalized socially or politically, or have greater difficulty in accessing treatment and care. This often includes infants and lone children, pregnant women, older people, people with disabilities, minority ethnicities, and geographically isolated groups.

Table 1.2 **Population displacement and risk of malaria in resettlement areas**

<table>
<tr><th colspan="2">Population movement</th><th rowspan="2">Risk</th></tr>
<tr><th>FROM area with:</th><th>TO area with (or through area with):</th></tr>
<tr><td rowspan="2">No or little transmission</td><td>High transmission</td><td>All age group displaced persons at risk (expected malaria outbreak)</td></tr>
<tr><td>No or little transmission</td><td>Not at increased risk</td></tr>
<tr><td rowspan="2">High transmission</td><td>High transmission</td><td>Children and pregnant women at greatest risk. Also risk may be increased by stress of displacement, or by increased exposure to vectors.</td></tr>
<tr><td>No or little transmission</td><td>Possible explosive outbreaks in the non-immune host population depending on the malaria receptivity status (possibility of transmission in area)</td></tr>
<tr><td>Returnees from settlement area with low transmission</td><td>Area of origin with high transmission</td><td>If return after a period of one year, assume that returnees have lost most of their semi-immune status. Children born outside their area of origin will be especially at risk.</td></tr>
</table>

Effective humanitarian response

An effective humanitarian response relies on many factors. Six principles for effective emergency response are listed below. Each is described more fully in relation to malaria control in the relevant chapters of this publication:

- **coordination** of responses, advocacy, and resource mobilization (Chapter 2)
- **accurate and timely assessments** of the situation and those affected (Chapter 3)
- **operational planning** of the response (Chapter 3)
- **implementation** of the response and specific interventions (Chapters 4–8)
- **operational research** to improve intervention delivery (Chapter 9)
- **monitoring and evaluation** to support coordination and planning (Chapters 4–5).

References

- ALNAP (2003). ALNAP *Annual Review 2003: Humanitarian Action: Improving monitoring to enhance accountability and learning.*
- IASC (1994). *Working Paper on the Definition of Complex Emergencies.* IASC Secretariat, December 1994.
- UNISDR (2004). *Living with Risk: A global review of disaster reduction initiatives.* United Nations Inter-Agency Secretariat of the International Strategy for Disaster Reduction.
- WHO. (2012). *World Malaria Report 2012.* Geneva, World Health Organization. http://www.who.int/malaria/publications/world_malaria_report_2012/en/

Finding out more

- Howard N, Sondorp E, and A Ter Veen, eds (2012). *Conflict & Health.* Open University Press.
- Warrell D and H Gilles, eds (2002). *Essential Malariology.* 4th ed. London, Arnold.
- WHO (2009). *Health Cluster Guide: A practical guide for country-level implementation of the Health Cluster* (Provisional version – June 2009). Geneva, World Health Organization.

CHAPTER 2
Coordination

This chapter:

- outlines key coordination mechanisms for preparation and response in humanitarian emergencies
- summarizes advocacy and resource mobilization considerations for malaria control in humanitarian emergencies
- describes principles, priorities, and constraints during preparation and response

Coordination mechanisms

The cluster approach

Since 2005, the cluster approach has been used, both globally and at country level, to plan and organize the international response to humanitarian emergencies. This approach aims to improve the predictability, effectiveness, and accountability of humanitarian responses by grouping humanitarian actors (e.g. UN agencies, national and international NGOs, the International Red Cross and Red Crescent Movement, and civil society) into sectoral groups called clusters. Each cluster is headed by a cluster lead agency whose role is to facilitate a coordinated response and to support national capacity as much as possible. Refugee settings are an exception, as the cluster approach is not activated, and UNHCR is the lead agency. For new humanitarian emergencies the cluster activation procedure entails the following:

- The UN humanitarian coordinator (HC) or resident coordinator (RC) consults relevant partners;
- S/he proposes leads for each major area and sends a proposal to the Emergency Response Coordinator (ERC);
- ERC shares the proposal with Global Cluster Leads;
- ERC ensures agreement at global level and communicates agreement to HC/RC and partners within 24 hours;
- HC/RC informs the host government and all partners.

There are currently eleven clusters; malaria control falls under the Health Cluster, which is led by the World Health Organization (WHO). Additionally, the Inter-Agency Standing Committee (IASC), established in 1992, is the primary mechanism for inter-agency coordination of humanitarian assistance.

Coordination among partners

Effective malaria coordination is multisectoral, both globally and at country level. Malaria control will include clusters and key sectors such as public health, water and sanitation, shelter, and social protection/community services. Operational partnerships are essential among a range of organizations. It is critical to work with national and local government agencies when planning and implementing an emergency malaria control response. Existing health facilities and national staff play an important role in any response and, with international support, are often best placed to deliver emergency health care.

Local NGOs, faith-based organizations, and community groups are important partners in an emergency response, particularly after the acute phase. Local and refugee communities have important skills, influence and cultural understanding, which may be lacking in the international humanitarian community, and it is important to identify existing capacity as soon as possible and work together with affected communities.

Advocacy and resource mobilization

Global and local advocacy considerations

Universal access to services for malaria control is a declared international goal, promoted by WHO, the Roll Back Malaria (RBM) Partnership and others. For example, Millennium Development Goal 6 includes reducing malaria incidence among its targets. Inclusion of conflict-affected and displaced populations is necessary if the world is to achieve this global target.

Access to malaria control programmes is an important public health priority for populations in malarious areas affected by humanitarian emergencies. Integration of these populations into national malaria control programmes and strategies provides important benefits, both for affected populations and for global and national malaria reduction targets. To ensure this inclusion, strong and concerted advocacy at global, regional, and country levels is necessary. Improved coordination among governments, the UN system and civil society during the planning and revision of national strategic plans is needed.

In addition to ensuring emergency planning in national malaria control strategies, malaria control should be included in humanitarian emergency preparedness and contingency planning at country level. This inclusion will not only allow an improved coordinated response, but will provide the flexibility to prioritize and rapidly transfer the necessary malaria control funds from routine to humanitarian programming.

Resource mobilization

There are four key ways of obtaining funds for malaria control needs in humanitarian emergencies:

1. The inclusion of malaria control in humanitarian planning and appeals, such as flash and consolidated appeals

A flash appeal is launched five to seven days after the onset of an emergency in order to fundraise for the first three to six months. Malaria control activities during the acute phase of the emergency should be included in this appeal.

A consolidated appeal is a multisectoral document developed among agencies. The appeal is not only a fundraising document, but also serves as an important tool for the planning, coordination, implementation and monitoring of humanitarian activities.

Malaria control programmes should be included under the health section of the common humanitarian action plan (CHAP) with a coherent set of activities that correspond to agreed priorities and strategies.

2. The inclusion of malaria control in proposals for grants from the UN Central Emergency Response Fund (CERF)

The Central Emergency Response Fund (CERF) is a stand-by fund established by the UN to enable timely and reliable humanitarian assistance to victims of natural and human-induced disasters. Malaria control activities can usually be included in both the *Health in Emergencies* submission, for activities that have an immediate impact on the health of emergency-affected populations, and as part of the water, sanitation and hygiene (WASH) submission, when emphasising vector control.

3. Global Fund to Fight AIDS, Malaria and Tuberculosis

The primary source of funding for national malaria control programmes for many countries is the Global Fund to Fight AIDS, Tuberculosis and Malaria (Global Fund). Ten percent of allocated funds can be reprogrammed at country level and this should be explored if an emergency occurs in areas where funding has been designated. A good example of this was during the major floods in Pakistan in 2010.

The importance of including provisions for populations affected by humanitarian emergencies in strategic planning documents has been recognized in the recent decision by the board of the Global Fund to move towards funding national malaria strategic plans in its new funding model.

4. Assessing funds from other sources

A number of other potential funding sources exist. For example, funds can be sought from the World Bank, the Office of U.S. Foreign Disaster Assistance (OFDA), Multi-Donor Trust Funds, and through the United Nations Development Assistance Framework (UNDAF).

Priorities and constraints

The priority in a humanitarian emergency is to reduce excess mortality and morbidity due to malaria as quickly as possible. Unsustainable approaches (e.g. augmenting existing infrastructure with temporary expatriate staff and/or parallel facilities) can be justified during the acute phase. However, a number of general constraints are encountered in humanitarian emergencies, some of which are listed in Table 2.1.

Finding out more

- CERF Life Saving Criteria and Sectoral Activities http://www.who.int/hac/network/interagency/news/cerf/en
- Inter-Agency Standing Committee (IASC). (2006). *Guidance note on using the cluster approach to strengthen humanitarian response*. Geneva, IASC.
- Inter-Agency Standing Committee (IASC), WHO (2009). Health Cluster Guide. *A practical guide for country-level implementation of the Health Cluster*. Geneva, World Health Organization.
- Spiegel et al. (2010). Conflict-affected displaced persons need to benefit more from HIV and malaria national strategic plans and Global Fund grants. *Conflict and Health* 2010, 4:2.
- The Global Fund. http://www.theglobalfund.org/en/

Table 2.1 **Constraints in humanitarian emergencies**

Health infrastructure
Local health facilities in humanitarian emergencies are often in poor condition, as affected areas have frequently experienced many years of social and economic decline before the onset of conflict or political strife. Infrastructure may be destroyed during the emergency. Modern intensive care facilities and equipment, and even relatively basic inpatient equipment and supplies such as X-ray and oxygen, are rarely available or, if available, cannot be maintained in emergency conditions.
Human resource capacity
During the acute phase of many emergencies, national capacity may be weakened and civil structures, including health and other services, are often unable to cope with the scale of response required. Shortages of skilled staff are the norm. Skilled national professionals, who have the means to relocate from the affected area, are often the first to leave in conflict situations, further weakening national capacity. There is often a lack of the specific technical skills required to ensure health care delivery in difficult and challenging situations. Capacity may remain weak for many years.
Local transmission data
During the acute phase of emergencies, there is often confusion or lack of information about the presence and types of vectors and about the potential for local transmission in temporary settlements. This has implications for prioritization of preventive services in addition to diagnostic and curative services.
Logistics
Supply, storage and distribution problems are often exacerbated by poor road infrastructure, looting of stores, and breakdown of procurement systems and supply management.
Security
Poor security complicates logistics and service provision, restricts the access of vulnerable populations to health care, and places emergency teams under great stress, resulting in high staff turnover.
Funding
Delayed funding decisions, insufficient funding, and a rapid decline in funding availability are key problems. Resources for training field teams in malaria control methods may also be severely limited.
Coordination
Lack of transparency, complementarity, and information sharing among partners can hinder effective response.

CHAPTER 3

Assessment and operational planning

This chapter:

- outlines essential and desirable information for analysing malaria burden and risk in humanitarian emergencies
- describes appropriate data collection activities if such data is not already available
- explains operational planning for an effective malaria response

Assessment and information needs

Assessment in a humanitarian emergency context is used primarily to determine the level of malaria risk and the capacity to respond. The following guiding principles can be used for the assessment, planning and selection of malaria control activities:

- maximize the use of existing information at international, national, district and community levels;
- carry out rapid surveys if existing information is inadequate or inaccessible;
- link malaria control interventions to current effective national policies;
- use available local expertise to assist with the selection of malaria control options;
- involve affected populations in decision-making and action.

Essential and desirable information

In an emergency, information about the demographics of affected populations, local malaria parasites, vectors, malaria endemicity, transmission, and response capacity, is critical in planning and implementing control measures. General information is needed to:

- identify current health priorities and potential health threats;
- assess the capacity and resources available to respond;
- collect baseline information for monitoring and evaluating the effectiveness of planned interventions.

Malaria-specific information is needed to:

- determine whether malaria is a major health problem (e.g. actual malaria disease and death burden in host and displaced populations);
- determine if there is local transmission and whether the area and context increase the likelihood of malaria outbreaks (e.g. immune status of populations, geographical and climatic conditions, environmental changes, local malaria parasite species, local vector species and their behaviour, population knowledge and behaviours);
- determine which population groups are most at risk of malaria infection, illness and death (e.g. ages, gender, and socioeconomic groups most affected) so that service delivery can be prioritized;
- identify any constraints to implementing an effective response (e.g. inadequate health policies, weak health system structures and capacity, limited access to health services, and inability to mobilize international and national partner support);
- ensure pesticide regulatory issues are addressed: LLINs, and insecticides for IRS and for larviciding, where appropriate, should be WHOPES-recommended and may need to be registered by the national regulatory agency before use; insecticide-treated materials, such as tarpaulins and blankets, may require agreement by national authorities prior to in-country usage; and some relief agencies will need to obtain an environmental assessment or waiver before using vector control products that are not formally recommended by WHO. WHO will consider vector control products for policy recommendation on an individual basis as data for those products become available.

Sources of information

Assessments should be conducted jointly with national authorities and key partners, and results shared with international and national partners so they can contribute to overall assessment of the health situation and wider humanitarian needs. The initial situation assessment should take less than a week, although this will depend on security, transport, and communication.

Basic information on malaria in the country can often be found in national malaria strategic plans, reports of recent malaria programme reviews, the WHO annual World Malaria Reports, Roll Back Malaria Partnership Country Facts, country-specific Malaria Operational Plans of the President's Malaria Initiative (PMI), Global Fund country grant portfolios, contingency information and vector ecology profiles from the US Armed Forces Pest Management Board (http://www.afpmb.org/), and in information on national registration status and vector-borne disease data from major WHOPES-approved pesticide manufacturers.

Where malaria information relevant to the emergency zone is limited or outdated, particularly where government structures and services have been severely disrupted, it may be necessary to carry out rapid surveys to assess the situation and the relative importance of malaria. Surveys can include clinic-based fever surveys or cross-sectional prevalence surveys (see the *Rapid epidemiological surveys* section below and Annex III).

Data collection

Information on context and demographics

Use the general checklists below to ensure relevant information is gathered. Data collection should include information on:

- the background to the emergency situation;
- transport, roads and infrastructure, including access to airports;
- security;
- communications;
- environments and climate;
- international and national partner locations and activities;
- coordination mechanisms.

Data collection on the demographics of affected populations should include:

- estimated total population size and breakdown by age and gender (see Chapter 4);
- estimated mortality per 10 000 per day (see Chapter 4);
- origins of displaced or refugee populations;
- other known or measurable health indicators (e.g. prevalence of malnutrition, previous exposure to malaria);
- distribution and settlement patterns;
- ongoing movement or mobility;
- type and location of dwellings;
- main professions;
- gender roles;
- main languages spoken.

Information on malaria transmission

Data collection should include:

- previous exposure of displaced and local populations to malaria (see Figure 3.1) – this will help in determining which populations may be most at risk;
- where possible, information on elevation, potential breeding sites, and seasonality of risk (e.g. climate, season, rainfall, temperature).

Information on malaria morbidity

Data collection should include information concerning the malaria burden in affected populations and on which population groups are most affected.

Figure 3.1 **Assessing malaria risk**

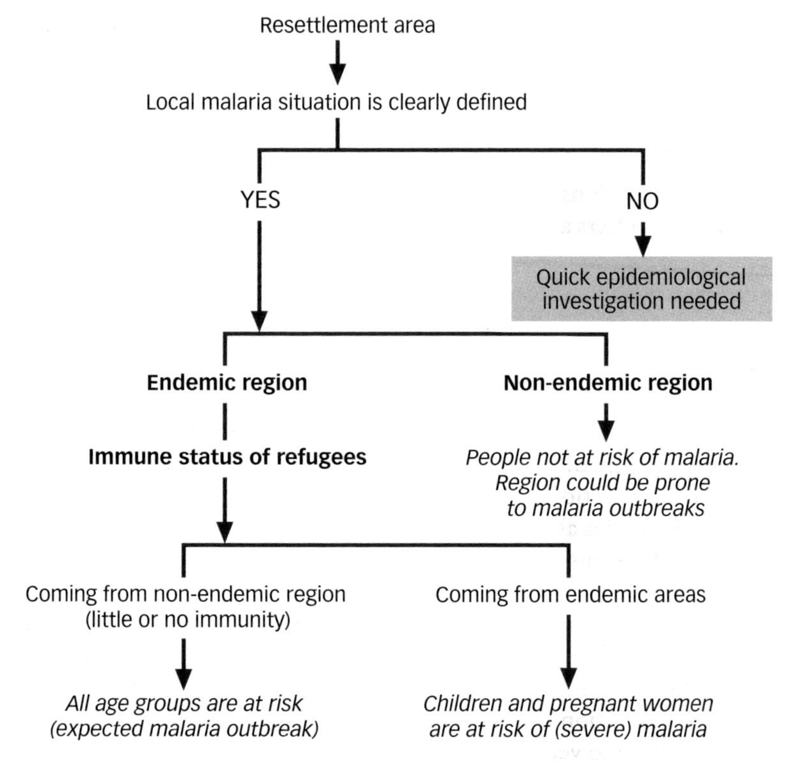

Source: WHO. (2005). *Malaria control in complex emergencies: An inter-agency field handbook* (First Edition). Geneva.

As a minimum, both clinical symptoms (e.g. actual or reported fever) and the presence of malaria parasites in those with fever should be measured (by RDT or microsocopy). This information can often be obtained from provincial, national, district and local levels through:

- a review of past and present medical records from health facilities in affected locations (e.g. number of fever or suspected malaria cases and confirmed malaria cases stratified by age group [usually ≤5 years and >5 years of age], and severe anaemia data [usually haemoglobin <77 g/dl] if available). Malaria disease trends can be assessed from a review of surveillance data over time;
- a rapid parasite prevalence survey at a health facility or mobile clinic, preferably one that sees patients from all affected groups (i.e. host, refugee, and internally displaced). This will indicate the actual proportion of confirmed malaria cases among all people presenting with fever (see *Rapid epidemiological surveys* section below and Annex III).

Information on malaria mortality

Specific mortality data are often unavailable or inaccurate. Data collection should include:

- malaria admissions and deaths, disaggregated by age and sex (both in absolute numbers and as a proportion of all admissions and deaths).

Data should be obtained from referral hospitals or facilities that admit patients with severe disease, with the caveat that such data usually represent only a fraction of overall malaria mortality at community level.

Data on malaria parasite species present in the area, and the relative importance of each, will affect the choice of diagnostic test, antimalarials procured for treatment, and possibly the choice of control measures. This information can be obtained from laboratory records or, for greater accuracy, from a cross-sectional parasitological survey, when time, resources, and operational realities allow.

Data on vector susceptibility to insecticides is paramount. Data on local vectors, their behaviours, and location of major breeding sites is also useful for understanding transmission patterns and choosing the most cost-effective vector control options (see Chapter 7). This information is particularly useful to avoid placing settlements in high-risk areas and to select potential additional prevention measures. The following information should be collected in relation to vectors:

- Species
- Susceptibility to insecticides
- Preferred breeding sites
- Feeding habits (e.g. time, location, host preference – usually inferred from species identification)
- Resting habits (e.g. inside, outside)
- Seasonal density changes
- Time and location that people come into contact with vectors.

Information on local vectors is often difficult to obtain but can frequently be gathered from ministry of health and national malaria control programme records or staff, records from previous entomological surveys (e.g. agency reports or "grey" literature maintained on websites such as WHO and the Armed Forces Pest Management Board [AFPMB]), and published literature.

Rapid epidemiological surveys

Rapid surveys may be used when the local malaria situation is not clearly defined (see Figure 3.1). They are used to assess whether malaria is, or may become, a significant problem, and to monitor trends. Rapid malaria surveys can:

- identify population groups most at risk of malaria (as this will not always be pregnant women and children under the age of five);
- estimate the proportion of the population, both symptomatic and asymptomatic, infected with *P. falciparum* and other malaria species; and
- help determine priorities for action, including the most appropriate case management and vector control measures.

Rapid surveys assess basic clinical signs and symptoms in a patient and parasite presence using microscopy or RDTs. Two types of rapid malaria surveys are:

- clinic-based among patients attending a health facility or mobile clinic;
- cross-sectional prevalence survey across the affected population.

Details on conducting these surveys are provided in Annex III. Conducting malaria surveys at the same time as other rapid surveys (e.g. population nutritional status, vaccination coverage) could save time and money.

Information on malaria outbreaks and control efforts

Data collection should include:

- presence of known risk factors for malaria outbreaks and whether the area has experienced earlier outbreaks;
- risk factors including large influxes of non-immune people into an area of high transmission, rainfall after an unusual period of drought, recent environmental changes that have increased suitable breeding sites, outdated treatment protocols, or drug and/or insecticide resistance;
- if there have been earlier outbreaks, a review of the number of cases and deaths and the causes of the outbreak;
- a review of past and present malaria control efforts in the area.

Information on population knowledge and practices

Data collection should include:

- knowledge, attitudes and beliefs among different ethnic groups about malaria and its causes, prevention and treatment;
- current usage of protective measures such as insecticide-treated materials, types of shelter and, where possible, family sleeping arrangements;
- treatment-seeking behaviour and health care provision by traditional healers and private practitioners;
- other practices (e.g. night-time behaviours) that could increase exposure to vectors.

Assessing capacity to respond

Response capacity in an emergency will depend on existing national policies, health system and services, population knowledge and practices, and the capacities of international and national stakeholders. Assessing response capacity in a humanitarian emergency requires information on existing health policy and services, infrastructure, relevant programmes, drugs and insecticides, and service usage.

Information on health policy, planning, and services

Data collection should include:

- current health policy (e.g. health sector reforms);
- financial mechanisms (e.g. user fees);
- responsibilities for planning and implementation at national, regional, provincial and district levels.

Information on health infrastructure and personnel

The actual condition of public and private health facilities and other infrastructure should be assessed, as should the availability of personnel, particularly those who have worked in malaria control. Assessment should include:

- types of health facilities, including those set up by NGOs and mobile structures;
- numbers of health workers and their level of training;
- staff turnover, especially for international staff;
- clinical algorithms for diagnosis, including Integrated Management of Childhood Illness (IMCI) strategy usage;
- treatments protocols for uncomplicated and severe malaria;
- availability of diagnostic testing (e.g. microscopy, RDTs);
- referral services;
- drugs and equipment, including blood transfusion facilities;
- health service logistics (e.g. transport facilities for patient transfer, effectiveness of communication between different levels of the system);
- surveillance, monitoring, and information management systems;
- supervision mechanisms.

Information on the National Malaria Control Programme

Data collection should include:

- programme goals, objectives, strategies, targets and action plans;
- programme structure, management and staff;
- protocols and guidelines, including those on dealing with malaria in emergencies and during epidemics.

Information on other priority health programmes and activities

Data collection should include:

- the national essential drugs list;
- drug stocks, supply and distribution systems;
- environmental health activities (e.g. vector control, insecticide stocks and supply lines);
- mother and child health initiatives, such as Making Pregnancy Safer (MPS), and IMCI.

Information on drugs and insecticides

National policies and regulations govern the use of drugs and insecticides in all countries, so it is important to collect information on:

- malaria diagnostic testing policy and practices;
- antimalarial drugs currently recommended by the ministry of health for treatment of uncomplicated malaria, management of severe malaria, and other drugs used locally (see Chapter 6);
- potential emergency drug options that are registered or whether these will need to be negotiated and introduced;
- which insecticides are recommended and used (see Chapter 7);
- available evidence on the efficacy of drugs and insecticides being used under national protocols;
- capacity for quality control of drugs and insecticides imported for the emergency;
- regulatory policies, including import and customs regulations and excises.

Information on accessibility and use of health services

Information should be collected on:

- host and displaced populations' access to, and use of, existing health care services, and the extent to which these services could be strengthened;
- factors likely to reduce access and treatment-seeking behaviour significantly in emergencies (e.g. poor security, culturally inappropriate services, preferences for alternative sources of health care, full or partial cost recovery mechanisms for consultation, diagnosis and treatment). Planning should include working with private treatment providers (if these are commonly used by the population that will be served).

Operational planning

Multisectoral planning

The capacity to respond to a humanitarian emergency will often depend on international and national partners. It is therefore important once the presence and activities of other agencies has been assessed, to determine mechanisms and scope for coordination and joint planning. As malaria control goes beyond the health sector, it is vital that multisectoral planning is undertaken based on the malaria risk and response capacity assessments.

Site planning

In the initial phase of humanitarian emergencies, refugees and internally displaced people move quickly to the nearest safe place outside the conflict area. Subsequently, international organizations and the host government may resettle refugee or internally displaced populations in more permanent sites. Whether or not this happens depends on the local situation. For example, some local authorities may be reluctant to encourage the establishment of long-term settlements.

If displaced populations are to be resettled, criteria for site selection include the availability of water and land, means of transport, access to fuel, and security (and possibly distance from borders). However, areas that meet these criteria are usually already occupied by resident populations, so displaced populations are instead often resettled in less desirable sites. If there is a choice when selecting a site, it is helpful to consider specific criteria that may affect the risk of malaria:

- assess the epidemiological characteristics of the area;
- investigate potential vector breeding sites, making use of local expertise and knowledge of important vectors;
- avoid sites close to major breeding locations of local vectors. In Africa, such sites often include marshy areas and flat, low-lying land at risk of flooding. If possible, sloping, well-drained sites with tree cover and sheltered from strong winds, should be chosen. In South-East Asia, thick forest should be avoided, as such areas are frequently a suitable habitat for efficient malaria vectors.

Selecting vector control activities

If malaria transmission is very high, rapid prescriptive standardized approaches to vector control may be necessary, such as long-lasting insecticidal nets (LLINs), indoor residual spraying (IRS), or provision of blankets or other materials that have been treated with insecticide (see Chapter 7). However, costly and logistically intensive activities such as IRS require a clear understanding of vector behaviour and the susceptibility of local vectors to available insecticides. In addition, understanding population knowledge and practice is critical in selecting appropriate vector control activities (see Chapter 8).

Selecting effective first-line antimalarial treatment

Since early testing of suspected malaria cases and prompt effective treatment of these are the first priority in malaria control, it is essential to select and use a first-line antimalarial drug that is efficacious and safe, with limited contraindications and side-effects (see Chapter 6).

Managing supplies of diagnostics, antimalarials, and other essential medicines

After diagnostic tests and antimalarial medicines have been selected, important steps in establishing systems to manage supplies include:

- establishment of a working group or interagency committee, involving local and international agencies and organizations, to manage the emergency;
- establishment of a working group subcommittee to manage supplies of diagnostic tests and essential medicines, including antimalarials. The agency with the most experience and expertise in supply management of essential medical commodities should take the lead, with responsibility for receiving, storing and distributing goods and monitoring the supply system. It is important to involve the relevant health authorities in the host country;
- careful attention to quality of diagnostics and medicines and sources of supply. It is important to check product specifications, including diagnostic performance, expiry dates, and in the case of medicines, dosage form and strengths;
- development of a stock management system that is as simple as possible and within the scope of the host country (e.g. storage conditions for medicines and diagnostic supplies in relatively cool and non-humid conditions is important to prevent deterioration);
- development of a clear policy for free distribution of diagnostic testing and medicines, ensuring that all agencies and communities are aware of this policy;
- ensuring that all staff who provide health care are trained in rational use of medicines, including the use of diagnostic testing and adherence to recommended antimalarials. Training should be arranged if necessary;
- seeking advice from the national health authorities if donated supplies of medicines are not approved or aligned with national treatment protocols. If there is no other option, health workers should be trained in the rational use of these medicines as a temporary measure to save lives, while the procurement of approved medicines is started as soon as possible;
- ensuring patient counselling to improve adherence to treatment and advise on how to recognize symptoms that would require a return to the

clinic. This may not be feasible in the acute phase of an emergency where there is a heavy case load, but should be possible once the displaced or refugee population is living in stable settlements and during the chronic phase;
- an early decision, with health authorities, on whether or not the resident population will also have access to health care services delivered as part of humanitarian assistance. If so, estimates of requirements for the resident population must be included when diagnostic test and drug supply requirements are calculated.

Procurement and supply chain management

Procurement and supply chain management (PSM) are crucial to ensuring that interventions reach affected populations. Providing supplies quickly and cost-effectively is often a great challenge. It is important to be aware that:

- countries need to have prepositioned emergency stocks;
- the main procurement bodies for malaria intervention commodities are governments, United Nations agencies and international organizations. Different partners support countries with product procurement;
- PSM for malaria is usually integrated with other diseases. Interagency Emergency Health Kits (IEHK) are standardized to provide for the basic health needs of 10 000 people for three months while health services are disrupted. There is a specific malaria module for emergencies occurring in malaria endemic areas, which includes RDTs, artemether-lumefantrine, and injectable artesunate to treat falciparum malaria;
- supply storage and distribution can be challenging due to damage to buildings and road infrastructure, to shortage of vehicles, and to looting of warehouses or lack of storage space. A number of malaria interventions, such as IRS and LLINs, require significant storage space.

Community outreach

Community participation and effective health education are essential to the success of malaria control interventions in humanitarian emergencies and attention must be paid to understanding community perspectives and the sociocultural factors that influence community behaviours (see Chapter 8).

Finding out more

- IASC (2006). *Guidance note on using the cluster approach to strengthen humanitarian response*. Geneva, Inter-Agency Standing Committee
- Inter-Agency Standing Committee (IASC), WHO (2009). *Health Cluster Guide: A practical guide for country-level implementation of the Health Cluster*. Geneva, World Health Organization.
- Spiegel et al. (2010). Conflict-affected displaced persons need to benefit more from HIV and malaria national strategic plans and Global Fund grants. *Conflict and Health*, 4:2
- WHO. Information note on recommended selection criteria for procurement of malaria rapid diagnostic tests (RDTs). Available at http://www.who.int/malaria/publications/atoz/9789241501125/en/index.html
- Test.Treat.Track Initiative. See details at http://www.who.int/malaria/areas/test_treat_track/en/index.html
- Guidelines on procurement of public health pesticides. See http://www.who.int/malaria/publications/atoz/9789241503426_pesticides/en/index.html

CHAPTER 4
Surveillance

This chapter:

- defines surveillance for health outcomes, including malaria, in humanitarian emergencies
- outlines the basic information needed to plan and implement malaria surveillance in humanitarian emergencies, including what data to collect, how to collect data, and how to use data
- describes the use of surveillance for monitoring and evaluation in humanitarian emergencies

Health surveillance in humanitarian emergencies

Definition

Surveillance for health outcomes in humanitarian emergencies, as in other settings, refers to the systematic collection, consolidation, and evaluation of relevant health data, as well as the prompt dissemination of results to those who need them. For example, malaria surveillance may be used to: identify malaria as a health problem; monitor trends in its occurrence; give early warning of potential malaria outbreaks; and monitor the effectiveness of control interventions. Surveillance for health outcomes, including malaria, is essential for planning, implementing and evaluating health interventions. In a humanitarian emergency, a surveillance system is a priority that must be established quickly. A well-functioning surveillance system can help determine which conditions, including malaria, are operational priorities. Malaria surveillance should itself be integrated into an overall surveillance system that collects data on all potentially significant causes of morbidity and mortality.

Surveillance approaches

In humanitarian emergencies, surveillance is most often based on data collected by health care providers and reported at regular intervals through health facilities (e.g. clinics, hospitals). Where necessary this data can be complemented by active surveillance or periodic surveys. Active surveillance includes an ongoing collection of information that is obtained either by periodically contacting health care providers who are not participating in routine surveillance, or via regular home visits. Periodic surveys are time-limited exercises that are usually conducted at the community level and are often carried out early in a crisis when routine surveillance systems are not yet established or are functioning poorly. Surveys can be repeated at intervals, ideally using the same methodology each time, to allow for the comparison of results as the situation progresses.

If no surveillance system is in place or the existing system is inadequate, it is essential to establish a functional system. This must, however, be carried out in a coordinated and integrated manner (see Box 4.1). Challenges to establishing an effective surveillance system in a humanitarian emergency may include:

- inadequate understanding at field level of what a surveillance system is and why it is needed, resulting in poor recording, reporting and use of information;
- poor motivation of health workers;
- insufficient diagnostic capacity to confirm clinical diagnoses, resulting in inaccurate and unreliable data;
- unknown representativeness of health facility-based data because only a small percentage of the population use health services (e.g. security

Box 4.1 **Surveillance challenges after the Haiti earthquake, 2010**

Following the 2010 Haiti earthquake, over one million people were left homeless, with the majority of the displaced population living outdoors or in temporary shelters, putting them at increased risk of malaria. Throughout much of the earthquake-affected zone there was extensive destruction of health care facilities and of public health infrastructure, as well as the displacement and loss of many health care providers. After the earthquake, the chaotic and changing environment was a major challenge to efforts to set up malaria surveillance. Lack of coordination, logistical challenges, mobile populations, and a constantly changing network of health care facilities, made accurate and timely surveillance difficult. In this context, initial rapid active surveillance was conducted by mobile teams, who used RDTs for case confirmation and who provided locally effective treatment. The teams also relayed information back to the Haitian Government and its partners. As the situation became less chaotic, a more formal surveillance system was employed relying on a mixture of active and passive methods.

problems or cost can limit access for a proportion of the affected population);
- lack of coordination among agencies;
- restricted access to the population by health care workers and agencies (e.g. due to security issues).

Planning malaria surveillance during humanitarian emergencies

Collecting basic information

Before any surveillance system is implemented in a humanitarian emergency, it is necessary to have basic information about the size and structure of the population in question. In this chapter it is assumed that much of this basic information, such as estimated population size and demographics, has been collected by emergency response agencies and is available for use and incorporation into the health surveillance system.

In cases where basic information about the affected population has not yet been collected or is inaccessible, it can be collected following the methods described in Annex I. If such basic data is lacking, it is advisable to strengthen coordination between partners to ensure that as much information as possible is gathered as quickly as possible.

Population size and structure

Data on population size and structure is important for planning the scale and size of an emergency response as well as for allocating resources. It is also helpful for calculating health indicators, such as mortality and disease burden. Population size and structure can be determined using the methods summarized below and described in Annex I:

- counting households and the average number of people per household;
- mapping the site area and measuring average population density;
- conducting a census or reviewing records (e.g. of population, camp registration, food distribution);
- using information from programme activities (e.g. vaccine coverage).

Basic information on the number of deaths that have occurred (retrospective mortality) and data from verbal autopsies can also be of use and may already have been collected (see Annex I).

Surveillance priorities in humanitarian emergencies

In an emergency:

- collect only the information that is absolutely necessary;
- keep data collection as simple as possible;
- decide who will collect data, from where, and what training will be given;
- decide how often to collect data;
- budget for the costs of surveillance (see Box 4.2);
- analyse what the data reveals about trends and give feedback in relevant coordination forums and to data providers (e.g. clinics).

Box 4.2 **Budgeting for surveillance**

Surveillance should be seen as an activity in its own right and should, therefore, be specifically included in budgets for malaria control. Dedicated surveillance staff and activities should be developed.

Human resources

Usually a combination of local and health ministry resources (e.g. vector control teams from the national malaria control programme), possibly reinforced by external assistance and/or partners. Personnel required may include a surveillance coordinator or manager, an epidemiologist, and a laboratory technician. Outreach teams may also be needed for active surveillance.

Laboratory

Cost is dependent on the chosen diagnostic strategy (e.g. RDT, microscopy). Other laboratory equipment and associated consumables need to be included, as well as costs for quality control of laboratories and supervision of data collection (see Chapter 6).

Transport and logistics

Requirements and costs will depend on the terrain and on the accessibility of different areas, but adequate resourcing is crucial.

Training

Costs of training health workers and outreach teams should be included, and refresher training should be considered in most circumstances.

Other costs

Costs could include: the forms, equipment or materials required for analysing data; communication equipment; training and administration.

Data collection

In complex humanitarian emergencies, it is essential to prioritize surveillance of all-cause mortality, cause-specific mortality, syndrome-based morbidity, and disease-specific morbidity, including malaria. Where possible, this should be done in an integrated and standardized way through existing systems rather than in parallel to existing systems. Possible sources of data include:

- local community health workers and the community;
- health personnel working in hospitals (e.g. outpatient and inpatient departments), health centres, and mobile outreach teams;
- special teams trained to conduct surveys.

Mortality surveillance

Monitoring mortality is an essential component of a health surveillance system and overall mortality is important, even in a malaria-focused surveillance system. Death rates may be high in the acute phase of an emergency and the immediate priority is to reduce mortality as quickly as possible.

Mortality should be assessed across standard age groups and should distinguish deaths that occur among pregnant women. Indicators used for mortality surveillance include:

- number of deaths per 10 000 persons per day (crude mortality rate);
- number of deaths in children under five per 10 000 persons under five per day (under-5 crude death rate);
- disease and cause-specific mortality rates (e.g. malaria, measles, acute respiratory infections, diarrhoea, war-related violence or injuries);
- proportion of mortality attributable to specific diseases (see section below).

Ideally, crude death rate estimates would already be collected by emergency response agencies early in a humanitarian response and would be available for programmes aiming to control malaria. However, if a system has not been established, an active mortality surveillance system using home visits should be established and home visitors trained to record daily total deaths and deaths in children under 5 years of age. Daily data should be compiled and analysed at the end of each week (Figure 4.1). Surveillance for deaths that occur at facilities should be part of facility-based health surveillance. Community data from the active mortality surveillance system can be combined with health facilities mortality data, to ensure that all available data are included, but should be verified to avoid double-counting deaths

Figure 4.1 **Example: mortality surveillance reporting form**

Place: .... Reported by: ....

From:....../......./......... (day/month/year) To: ....../......./......... (day/month/year)

Population[a]

Population	End of previous week (*A*)	New arrivals	Departures	End of this week (*B*)	Average population (*A*+*B*)/2
<5 years old					
≥5 years old					
Pregnant women					

[a] Information about the population is needed to calculate mortality rate.

Mortality rate

	Number of deaths			Rate (deaths/10 000 per day)		
	Male	Female	Total	Male	Female	Total
<5 years old						
≥5 years old						
Pregnant women						

registered at the clinic and recorded in the community. Recording where a reported death occurred can help with verification.

Cause-specific mortality

In the mortality surveillance, information can be recorded about the cause of death (Figure 4.2). Possible data sources include health facility registers, home visits, and interviews with heads of households. However, data should be verified to avoid double-counting deaths registered at the clinic and recorded in the community.

Causes of morbidity

A standard case definition should be used for each disease, and morbidity should be assessed across standard age groups and among pregnant women. Indicators used for morbidity surveillance include:

- incidence rate (the number of new cases per week, expressed as a population percentage or per 1000, 10 000 or 100 000 persons);

Figure 4.2 **Example: cause-specific mortality reporting form**

Place: .... Reported by: ....

From:....../......./......... (day/month/year) To: ....../......./......... (day/month/year)

Deaths due to	**<5 years of age**			**≥5 years of age**			**Total**			**Percentage**
	M	F	Total	M	F	Total	M	F	Total	
Non-bloody diarrhoea										
Bloody diarrhoea										
Severe respiratory infections										
Suspected malaria[a]										
Confirmed malaria[b]										
Presumed malaria[c]										
Measles										
Malnutrition										
Suspected meningitis										
Injury										
Others										
Total										

[a] Suspected malaria (i.e. confirmed malaria + confirmed non-malaria + plus presumed malaria);
[b] Confirmed malaria is based on parasite-based diagnosis (e.g. RDT or microscopy);
[c] Presumed malaria is suspected malaria (as determined by clinical signs and symptoms only) that has been treated without testing.

- incidence rate by age-group or categories (e.g. the number of new cases in a specified time-period in under-fives, or in pregnant women).

In humanitarian emergencies, these data are often collected from daily records of fixed health facilities or mobile clinics providing outpatient and inpatient services to the population (Figure 4.3). Daily data should be recorded at the facility and collected and analysed at the end of each week by the central coordinating agency for surveillance. These data can also be collected actively using cross-sectional surveys. Care should be taken to avoid double-counting of patients who report illness and have been to a clinic. If active data collection is repeated it is important to maintain a standard data collection method to evaluate trends.

Figure 4.3 **Example: cause-specific morbidity reporting form**

Place: .... Reported by: ....

From:....../......./......... (day/month/year) To: ....../......./......... (day/month/year)

Diseases	**<5 years of age**			**≥5 years of age**			**Total**			**Percentage**
	M	F	Total	M	F	Total	M	F	Total	
Non-bloody diarrhoea										
Bloody diarrhoea										
Respiratory infections										
Suspected malaria[a]										
Suspected malaria tested[b]										
Confirmed uncomplicated malaria[c]										
Presumed uncomplicated malaria[d]										
Suspected severe malaria										
Suspected severe malaria tested										
Confirmed severe malaria										
Presumed severe malaria										
Malnutrition										
Suspected meningitis										
Injury										
Others										
Total										

[a] Suspected malaria (i.e. confirmed malaria + confirmed non-malaria + plus presumed malaria) is useful for calculating test positivity rate (see Table 4.1);
[b] Tracking the number of suspected malaria cases tested can be used to calculate test positivity rates;
[c] Confirmed malaria is based on parasite-based diagnosis (e.g. RDT or microscopy);
[d] Presumed malaria is suspected malaria (as determined by clinical signs and symptoms only) that has been treated without testing.

Malaria-specific surveillance needs

The purpose of malaria-specific surveillance is to:

- *Monitor trends*, e.g.:
  - — Are malaria cases or malaria deaths occurring and are they increasing or decreasing?
  - — Is the number of malaria patients as a proportion of the total number of patients rising?
  - — Is an increase due to new arrivals or to an increase in the transmission of malaria?
- *Provide early warning of an outbreak*, e.g.:
  - — Is malaria incidence increasing?
  - — Is the proportion of confirmed malaria cases (e.g. RDT positivity rate, slide positivity rate) increasing?
  - — Is the proportion of adults with malaria increasing?
- *Monitor effectiveness of malaria control interventions and, if necessary, redefine priorities*, e.g.:
  - — Do sick people access health care structures?
  - — What is the coverage and use of bednets?
  - — How effective is the detection and treatment of malaria cases?
- *Map cases*, e.g.:
  - — Is there a clustering of cases that could indicate a focus of local transmission or higher-risk activities?

Malaria surveillance should include the indicators summarized in Table 4.1.

Data collected should be based on the standard WHO case definitions (see Table 4.2). Use of standard case definitions and confirmation of cases are essential to ensure high quality surveillance with which to make informed judgements about progress in malaria control. Relying only on clinical diagnosis will provide inadequate data about malaria cases in all but the highest transmission areas and, as such, it is important to use a suitable confirmatory tool, such as RDTs or microscopy (see Chapter 6). Quality surveillance also depends on reliable and complete data reporting from health facilities and laboratories to a central data point; on adequate and timely analysis of data collected; and on appropriate dissemination to partners and stakeholders.

Information specific to malaria can be collected from health centre laboratories to provide information additional to the formal health surveillance system, including the number of tests (smear/RDT) performed and the number of positive tests (smear/RDT) – to estimate test positivity rates.

Table 4.1 **Useful indicators for malaria surveillance**

Rates	Calculation	Interpretation
Crude mortality rate (CMR)	Number of deaths/10 000 population per day	>1/10 000 indicates an emergency (a doubling of baseline mortality rate)
Malaria-specific mortality rate	Number of malaria deaths/10 000 population per day	Is malaria a health priority? Is malaria a significant proportion of overall CMR?
Proportion of mortality caused by malaria	Number of malaria deaths/total number of deaths	Is malaria contributing significantly to overall CMR
Malaria incidence rate	Number of new malaria cases/1000 population per week	Is malaria increasing? What action should be taken?
Case-fatality rate in all malaria cases	Number of confirmed malaria deaths/1000 confirmed malaria cases	Are cases being managed effectively?
Case-fatality rate in severe malaria cases	Number of severe malaria deaths/Total number of severe malaria cases	How effective is the referral service? How well are severe cases managed at community, clinic or hospital level?
Malaria test positivity rate	Number of positive RDTs or slides/Total RDTs or slides examined	What proportion of suspected malaria is actually malaria? An increasing rate may indicate the beginning of a malaria epidemic or a seasonal peak (see chapter 4).
Proportion of population (or target group) covered by the intervention	Number of intervention units (e.g. LLINs)/ Population of interest	Can intervention coverage be improved? Are targeted populations being adequately reached?

Where there is more than one species of malaria of concern (e.g. *P. falciparum* and *P. vivax*) it is advisable to note the species for each case if possible. Recommended treatments for these two species often differ. *P. falciparum* malaria is also associated with higher mortality than *P. vivax*, so the ratio of cases caused by each species is important for resource allocation.

Malaria incidence rates should be used to guide prioritization. Test positivity rates (e.g. RDT/microscopy) may be useful if health seeking, testing or reporting rates cannot be accounted for to assess changes in malaria incidence. These can be collected through an appropriate passive surveillance system from health facilities and other providers (e.g. CHWs). Surveillance data should therefore be used for prioritization and targeting of interven-

Table 4.2 **Standardized malaria case definitions**

Suspected malaria

Patient illness suspected by a health worker to be due to malaria. Criteria for suspected malaria usually include fever, but the precise criteria vary according to local circumstances. All patients with suspected malaria should be tested by either microscopy or a rapid diagnostic test (RDT).

Presumed (not tested) malaria

A suspected case in which the patient did not undergo a diagnostic test but was nevertheless treated for malaria. Such cases have been sometimes referred to as probable cases; however, in most settings, the chance that a suspected case will be confirmed is <50% and therefore the term probable is inappropriate and should not be used.

Confirmed malaria

A suspected case of malaria in which malaria parasites have been demonstrated, generally by microscopy or RDT. In some suspected cases with a positive test, particularly in populations that have acquired immunity to malaria, febrile illness may be due to other causes. Nevertheless, a diagnosis of confirmed malaria is still given. If concurrent disease is suspected, it should be further investigated and treated.

Confirmed severe malaria

A patient with laboratory confirmed acute malaria with signs of severity and/or evidence of vital organ dysfunction.

Note: Uncomplicated and severe malaria categories are intended to be mutually exclusive. For example, a patient who initially presents with uncomplicated malaria but then develops signs or symptoms of severe disease should be classified only as having severe malaria, and not counted twice. This also applies to situations where health services record cases as *suspected malaria* until diagnosis is confirmed by microscopy or RDT, after which the cases become *confirmed malaria*, or the patient is treated for malaria without testing, after which the case becomes *presumed malaria*. Presumed and confirmed malaria cases should be reported separately. Though confirmed and presumed cases are subsets of suspected malaria, it can be useful to track suspected malaria cases and the number of cases tested for malaria as well, in order to understand changes in confirmed and presumed malaria.

Presumed malaria death

Death of a patient who has been diagnosed with presumed severe malaria, i.e. treated for malaria without laboratory testing being performed.

Confirmed malaria death

Death of a patient who has been diagnosed with severe malaria, with laboratory confirmation of parasitaemia.

Source: WHO (2012) *Disease Surveillance for Malaria Control.*

tions. For example, during the Afghan refugee crisis in Pakistan, only those camps where incidence exceeded a certain threshold were targeted for vector control. Using this system, preventive interventions targeted the most endemic camps, making the interventions more efficient.

Establishing passive disease surveillance requires microscopy quality control. Technicians from a central reference laboratory can provide this service. These technicians can regularly monitor the technical accuracy of microscopists working in field laboratories and clinics. Malaria rapid diagnostic tests can also be used for routine diagnosis. Both RDTs and microscopy should be connected to a quality assurance system (see Chapter 6).

Monitoring and evaluation

Surveillance is an important programme component of a comprehensive monitoring and evaluation (M&E) plan. Monitoring, as distinguished from surveillance of health outcomes, is systematic collection and regular analysis of programme data (e.g. number of LLINs distributed per day or week), to determine day-to-day programme functioning and changes over time. It can be used for a variety of functions (e.g. directing resources, identifying bottlenecks, reporting to donors).

Evaluation is usually conducted at set points in the programme cycle to assess whether the programme has achieved its aims by measuring progress against specific indicators. Normally, multiple indicators are used. These can include input indicators (e.g. items procured on time), process indicators (e.g. number of health workers trained), output indicators (e.g. number of items distributed), or outcome indicators (e.g. proportion of population protected by intervention) and impact indicators (e.g. reductions in malaria incidence or malaria deaths).

Indicators are often agreed upon with donors before implementation and thus should be used as far as possible in M&E. Indicators should be designed using SMART or SMARTE**R** criteria (e.g. **S**pecific, **M**easurable, **A**greed, **R**elevant, **T**ime-bound, **E**thical, **R**ealistic).

Programme effectiveness indicators

Information on programme effectiveness should be collected and analysed every week to assess how well different components are functioning. Information that can be useful in monitoring programme functioning include:

- How many home visitors are working and where?
- How many patients are diagnosed and treated for malaria and where?
- How many health personnel are involved in programme activities?

- What kind of antimalarial drugs are available, and in which dosage strengths?
- How long is the delay between ordering drugs and receiving them at health facilities?
- Are there shortages of drugs? If so, how frequent are the shortages and how long do they last?
- Are rapid diagnostic tests available? Are there shortages of diagnostic supplies? If so, how frequent are the shortages and how long do they last?
- What percentage of febrile patients are tested using blood smears or RDTs in outreach activities or health facilities, and what is the delay (both average and extreme) before test results are known?

Monitoring **case management** helps demonstrate how effectively cases are being diagnosed and treated. An indicator that can be used to monitor case management is the case-fatality rate.

Monitoring **coverage** helps ensure that vulnerable groups are being reached or have access to services. Indicators that can be used to assess coverage include:

- proportion of sick patients with access to a health centre;
- proportion of affected geographical areas covered by the programme;
- proportion of pregnant women and children under five years of age covered by the programme;
- ownership/use of LLINs.

Other considerations when assessing indicators for surveillance

Trends in malaria cases and deaths may be influenced by changes in health care seeking behaviour, diagnostic effort at points of care, and completeness of reporting throughout the system. The following are some suggested indicators that can be used to monitor or evaluate the surveillance components of the malaria control programme:

- number of persons or proportion of the population seeking care through available points of care;
- number and proportion of suspected cases tested for malaria;
- proportion of facilities providing timely reporting of their weekly data;
- proportion of community workers providing timely reporting of their weekly data;
- proportion of community workers providing complete reporting of their weekly data;
- number of weekly or monthly analysis reports disseminated (and used in decision making);

- timely reporting at project or agency level;
- timely reporting towards broader platforms, for example, the lead coordinating agency, MoH, or other stakeholder meetings.

Surveillance for other aspects of malaria control

Surveillance for other aspects of malaria control, outside the purview of normal health programming in emergency situations, can also be conducted. The primary focus of malaria control in humanitarian emergencies is to save lives and prevent morbidity. As such, routine disease surveillance is a priority. Therefore, only if resources allow are the following malaria-focused surveillance activities worth considering:

Drug efficacy surveillance

If possible, and the drug-efficacy status of the area is unknown, it can be useful to identify at least one health facility site in which to conduct *in-vivo* antimalarial drug efficacy surveillance (see Chapter 8). However, unless changes in efficacy are expected to be extreme, sentinel surveillance for drug efficacy should not interfere with other priorities.

The WHO protocol for in-vivo efficacy surveillance recommends that clinical and parasitological parameters be monitored on days 0, 2, 3, 7, 14 and 28 (see Box 4.3). Efficacy surveillance allows the efficacy and safety of first-line treatment to be monitored and confirmed. Monitoring of drug resistance can also be done using *in vitro* tests that identify genetic markers of resistance (e.g. PCR). This technique is increasingly applied, though support of an expert laboratory is required.

It may also be important to monitor drug quality if poor effectiveness or efficacy has been detected or if locally manufactured drugs are applied and the quality is questionable. Such monitoring can be done more accurately by a specialist laboratory than by using off-the-shelf kits.

Vector surveillance

Knowledge of malaria vector composition and density is operationally useful. Differing vector habitats and behaviours may have an impact on which control measures are implemented. If no data on local vector composition is available (e.g. from the national malaria control programme, published papers, local universities), and local transmission has been determined epidemiologically, then vector surveillance is particularly important to determine: (i) what and where vectors are, and (ii) susceptibility status to insecticides. Vector surveillance and insecticide resistance monitoring

Box 4.3 **Classification of malaria treatment responses**

Early treatment failure (ETF)
- danger signs or severe malaria on day 1, 2 or 3, in the presence of parasitaemia;
- parasitaemia on day 2 higher than on day 0, irrespective of axillary temperature;
- parasitaemia on day 3 with axillary temperature ≥ 37.5 °C; and parasitaemia on day 3 ≥ 25% °C count on day 0

Late clinical failure (LCF)
- danger signs or severe malaria in the presence of parasitaemia on any day between day 4 and 28 (day 42) in patients who did not previously meet any of the criteria of early treatment failure; and
- presence of parasitaemia on any day between day 4 and day 28 (day 42) with axillary temperature ≥ 37.5 °C in patients who did not previously meet any of the criteria of early treatment failure.

Late parasitological failure (LPF)
- presence of parasitaemia on any day between day 7 and day 28 (day 42) with axillary temperature ≥ 37.5 °C in patients who did not previously meet any of the criteria of early treatment failure of late clinical failure.

Adequate clinical and parasitological response (ACPR)
- absence of parasitaemia on day 28 (day 42) irrespective of axillary temperature, in patients who did not previously meet any of the criteria of early treatment failure, late clinical failure or late parasitological failure.

Source: http://whqlibdoc.who.int/publications/2009/9789241597531_eng.pdf

requires specialist equipment (e.g. light traps) and laboratory infrastructure and should be conducted under the guidance of an experienced entomologist (see Chapter 7).

References

- WHO (2009). *Methods for surveillance of antimalarial drug efficacy*. Geneva, World Health Organization. Available at http://whqlibdoc.who.int/publications/2009/9789241597531_eng.pdf
- WHO (2012). *Disease surveillance for malaria control: an operational manual*. Geneva, World Health Organization. Available at http://www.who.int/malaria/surveillance_monitoring/operationalmanuals/en/index.html

Finding out more

- WHO Health Action in Crisis. See http://www.who.int/hac

CHAPTER 5

Outbreaks

This chapter:

- outlines malaria outbreak preparedness for humanitarian emergencies
- provides guidelines for investigating a malaria outbreak
- describes malaria outbreak response and follow-up

Outbreak preparedness

Epidemic malaria may occur in areas of normally low or seasonal malaria transmission where people have little immunity to the disease. It may occur in non-immune populations that have moved through or into endemic areas. It may also occur in areas that were previously malarious, but where control has reduced the case load over recent years while the area remains receptive to malaria transmission. In these cases, severe disease and deaths can occur in all age groups, with young children, pregnant women, malnourished individuals, and people with concurrent infections (e.g. HIV) the most vulnerable. Even when malaria transmission is not excessive compared with previous years, local epidemics with high mortality may occur among vulnerable displaced populations because of the concentration of people, lack of adequate housing for preventive measures resulting in increased exposure to mosquito bites, concurrent infections and malnutrition, and reduced access to effective treatment.

Determining epidemic risk

Outbreak preparedness and response is thus an important component of malaria control in humanitarian emergencies. However, in emergency settings, clearly defined malaria outbreak thresholds (e.g. mortality or morbidity rates) are rare and detection of outbreaks is not straightforward. Unusual or sudden increases in malaria mortality, proportional mortality or incidence rates, when compared with previous weeks or months, may suggest an outbreak. It is important to be aware that:

- malaria outbreaks may continue longer than other disease outbreaks during emergencies, lasting 3-4 months or even longer depending on climatic conditions;
- malaria outbreaks often have multiple underlying causes.

In stable situations, an epidemic threshold can be calculated using five or more years of historical data. On this basis, WHO has developed an epidemic threshold calculator based on historical data. However, in an emergency, it is unlikely that sufficient historical data will be available – unless the camp or settlement area has been established for several years and accurate malaria data has been collected, at least monthly and consistently, allowing comparisons with deviations from the mean or median.

In the absence of historical data but where recent weekly patient data are available, epidemic malaria should be suspected if there is an unexpected rise in the number of malaria cases, a rising malaria test positivity rate (i.e. increasing proportion of patients with fever who have confirmed Plasmodium infection) and/or a rise in the case-fatality rate – as non-immune people can die of malaria in a matter of hours, even in well-equipped facilities. Where no comparative data are available, a cluster of severe cases and deaths due to febrile disease warrants investigation.

Preparedness action plans

Measures for epidemic prevention and control can be implemented effectively only if they are supported by inter-agency coordination, an infrastructure of well-trained personnel, adequate supplies and equipment, supervision and evaluation. In areas and populations prone to epidemic malaria, it is important that partners agree and implement a preparedness action plan, at national and/or local-level, covering the following:

- *Supplies* – a sufficient stock of quality-assured essential laboratory diagnosis and treatment supplies for uncomplicated and severe malaria, including RDTs (see Chapter 6); bednets (LLINs); and equipment, materials, protective clothing and insecticide for emergency prevention activities (see Chapter 7).
- *Laboratory facilities* – an identified laboratory that can provide quality control, determine parasite species and density, and ensure safe blood transfusions.
- *Staff and transport* – staff and transport for mobile teams; capacity to expand inpatient and outpatient care facilities, including the number of skilled medical and technical staff needed to adequately respond to an

increased malaria case load; and capacity to improve transport and logistics for referral of severe cases to inpatient facilities.

- *Educational messages* – messages to encourage population uptake of preventive measures and early treatment-seeking, and materials and equipment to transmit these messages to the public.
- *Training and refresher training of staff* – on malaria diagnostic testing, antimalarial treatment, management of non-malarial febrile illnesses, vector control, weekly disease surveillance and survey methods.
- *Budget* – for all of the above.

Outbreak investigation

A suspected malaria outbreak should be investigated immediately as early intervention is critical. The purpose of an investigation is to confirm the need to scale-up malaria control interventions beyond normal implementation (e.g. extending clinic opening hours, intensifying vector control) and to differentiate between a genuine epidemic and normal seasonal variation. It is important to describe the outbreak using the person-place-time framework (e.g. patient characteristics, local context, changes in frequency over time). Formally declaring "an epidemic" may be of secondary concern given a sudden and unexpected rise in malaria burden. This is particularly true when no confirmatory historical data is available, as is often the case in humanitarian emergencies.

Confirming the outbreak

To confirm an outbreak, the first step is to visit the area, if it is accessible, and collect information from local health staff and community leaders. Ask if the number of malaria cases is unusual for the season. Have there been recent population movements that could account for the rise in case numbers? Have there been changes in mortality or in-patient admissions for malaria? This is also an opportunity to conduct a rapid needs assessment for diagnostic testing and treatment and to check if clinics are able to cope with a rise in suspected and/or confirmed malaria cases.

Assessment should determine whether there is evidence of increased morbidity and mortality, including among vulnerable groups, and look for evidence of increased transmission intensity and/or movement of non-immune people into a transmission area. Increased transmission intensity can result from increased vector breeding; the local area should be examined to identify potential new breeding sites (Box 5.1). One or more rapid prevalence surveys, such as test positivity rate among fever cases, should be included as part of this assessment (see Chapter 3).

Box 5.1 **Pakistan malaria outbreak**

In 2003, several thousand refugees crossed the border from Afghanistan into the tribal areas of Pakistan. They constructed mud-brick shelters, unintentionally creating vector breeding sites in the borrow pits from which they dug the mud. This resulted in high-intensity transmission of *P. falciparum* (Figure 5.1) in a relatively arid desert area and earlier in the season than would be expected. Note in Figure 5.1 that transmission took place earlier in the epidemic year and resulted in rapid rises in incidence over the course of only a few weeks.

Figure 5.1 **Epidemic malaria in Northwest Frontier Province Pakistan in 2003**, showing incidence of *P. falciparum* in a population of approximately 4000 Afghan refugees. Black lines indicate the refugee camp where the epidemic occurred, red lines indicate nearby camps and grey lines indicate the remaining six camps that had been established in the 1980s.

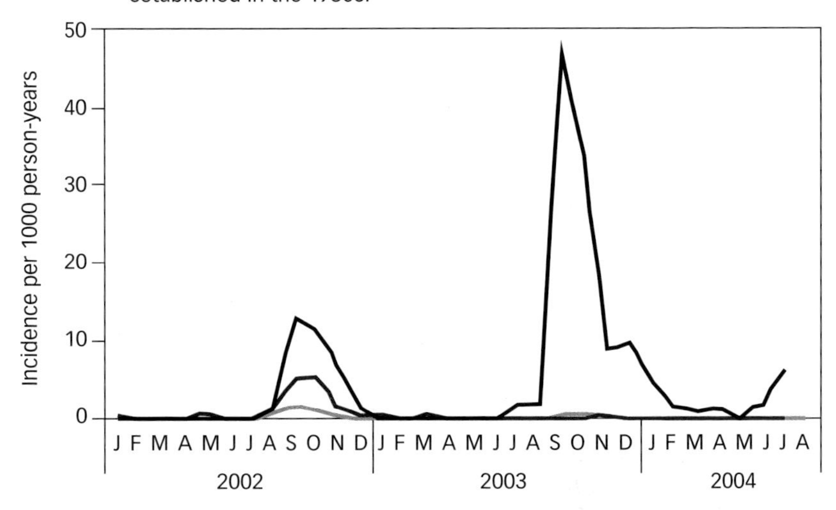

It is worth noting that malaria data can be misleading, and give rise to false alarms concerning potential outbreaks, particularly in the context of a changing health system. For example, an increase in case numbers resulting from improved malaria surveillance or removal of user-fees would result in increased numbers of detected and reported cases. This increase in cases, while real, is attributable to improved case detection rather than intensified malaria transmission. It is therefore important to collect as much information about local context as possible when investigating potential outbreaks, and to examine alternatives to the hypothesis that it is a true malaria epidemic.

Describing the outbreak

Planning an effective outbreak response requires a description of the outbreak. This should identify the timing and location of the outbreak to allow a focussed and measured response. In all settings it is important to rely on accurate malaria diagnostic testing (either RDTs or microscopy), and to gather data on confirmed malaria cases. Data on presumed malaria (i.e. clinically diagnosed malaria) may also be useful in the absence of other sources. The following outbreak description procedure is suggested:

- Discussion with local partners and those with knowledge of malaria transmission and local context;
- Analysis of information from rapid prevalence survey or surveys;
- Analysis of data collected through the surveillance system, including:
  - retrospective data from registers (if any);
  - prospective data collected using a malaria outbreak form (see Annex XI); the data should be stratified by population structure (age group, pregnant women) and geographical area (village, district, camp) so that the attack rate can be calculated for different groups and areas, and so that interventions can target the most vulnerable groups;
  - an epidemic curve, which shows graphically the number of new cases per week and the evolution of the outbreak. When regularly updated, the curve can show the transmission season, whether the epidemic has reached its peak, and whether preventive interventions are leading to a reduction in new cases (although variations in access to treatment may make interpretation difficult).

Outbreak response and follow-up

Coordinate planning and implementation

If an outbreak is confirmed, all health partners (e.g. government departments, UN agencies, humanitarian health organizations, and NGOs) need to:

- *Decide together how to tackle the outbreak* – Even if individual health actors will be limited to certain areas or population groups or focus only on certain methods (e.g. surveillance, disease management, vector control), it is essential that actions be coordinated to maximize the impact of interventions. Coordination is especially important if the response involves multiple layers of interventions (e.g. expanding inpatient capacity at referral facilities as well as improving access to first line care).

- *Choose agency focal-persons* – There should be a clearly defined focal-person for each organization and a lead-agency for central coordination if necessary (e.g. this is often WHO within the cluster approach).
- *Follow good planning principles* – Prioritizing services for those geographical areas and population groups that were identified as most-affected during the investigation phase.

Choose a strategy

The priority in a malaria epidemic is prompt and effective diagnostic testing and treatment with (in case of falciparum malaria) artemisinin-based combination therapy (ACT) in line with the national antimalarial drug policy. The operational strategy for disease management (e.g. whether to use mobile teams, additional or existing fixed units, or focus initially on improvement of inpatient or outpatient care) will depend on:

— the population affected and risk groups (during an outbreak, all age groups may be affected);
— the malaria proportional mortality rate;
— the case-fatality rate;
— the prevalence in different areas;
— access to existing health facilities
— distance to health facilities from affected areas.

Preventive measures and vector control should also be implemented if the situation warrants it (see Chapter 7). The provision of life-saving treatment for all malaria cases should be prioritized over vector control interventions if resources for the epidemic response are constrained. Annex V provides a checklist for effective malaria epidemic response.

Operational strategy components include ensuring treatment access, effective diagnosis, treatment, prevention of new cases, and response review.

Access to treatment

It is critical to ensure: that there are sufficient treatment points for the population to access treatment easily; that these are providing prompt and accurate diagnosis and effective case management with quality-assured ACT; and that vulnerable groups have access to treatment (see Chapter 6). Barriers to health services access (e.g. user-fees, poor organization of the

Box 5.2 **Outbreak response using mobile clinics in Burundi, 2009**

A rapid increase in malaria cases was identified in Burundi's northern provinces of Kayanza and Ngozi at the end of 2009. A field assessment, conducted by a team of Ministry of Health and Médecins Sans Frontières staff, identified overwhelmed clinics. Visits to villages showed that a significant proportion of people with fever had not consulted a health centre. The team decided to reinforce capacity in the hospitals and health centres, launch mobile clinics in villages and village clusters where high numbers of cases were reported, and target LLIN distribution. Fifteen mobile teams were deployed, with one team going three times a week to each site. Each team comprised four nurses, one driver, one person for registration and recording temperature, and one person to administer the first dose and explain treatment to the patient or parents. One nurse was responsible for triage, one conducted RDT diagnoses, and two took patient histories, conducted clinical examinations and dispensed drugs. Over five months, the mobile clinics contributed to covering a population of 537 000. The mobile clinics themselves identified 72 000 confirmed malaria cases and treated them with ACTs.

patient circuit, early closing times) should be removed to ensure that as many patients as possible have equitable and prompt access (see Box 5.2). Preventive treatment should be considered for pregnant women depending on the setting (see Chapter 7).

Active case detection can be conducted if resources are available. This involves parasite-based screening of fever cases at household level and treatment of all identified cases. If adequate diagnostic testing is not available, active case detection may, in rare instances, be based on fever screening only; in such circumstances, agreement and training on clinical case definitions is important in order to avoid inappropriate treatment of non-malarial illnesses. A potential approach for treating malaria infections that escape diagnosis is mass drug administration (MDA), the practice of treating a whole population within a given geographical area, irrespective of symptoms or diagnosis. Because of operational hurdles and safety concerns, and because its impact on transmission is generally short-lived, MDA should be considered only after careful consultation with malaria experts. The likelihood of increasing selection for drug-resistant genotypes through the use of MDA must also be considered.

Diagnostic testing

In all settings, including outbreaks, everyone with a positive malaria test result should be treated immediately, regardless of his or her symptoms.

In epidemic situations, where patient numbers are very high, RDTs can enable a team of two people to accurately screen up to 200 patients per day. Where this is not possible, because of shortages of either staff or RDTs, clinical diagnosis may be the only option. In such settings, once malaria has been established as the cause of the epidemic (e.g. through prevalence surveys) and agreement reached on a clinical case definition for malaria (also to avoid inappropriate treatment of non-malarial illnesses), presumptive treatment of fever cases with ACT is an appropriate strategy for reducing mortality, though only if systematic testing would overwhelm health facilities.

In a confirmed malaria epidemic, the proportion of fever cases presenting to health facilities is usually high and use of clinical diagnosis may be appropriate given personnel and/or supply limitations. If this approach is adopted, RDTs or microscopy should be performed on a proportion of clinically-diagnosed cases to track test positivity rate and the evolution of the epidemic. To ascertain whether the epidemic is continuing and to avoid over-treating, it is useful to carry out RDT surveys among patients in health centre waiting rooms at weekly intervals, or to make daily RDT checks of a percentage of consecutive fever patients. As the epidemic wanes, over-treatment will increase significantly if confirmatory diagnostic capacity is not established and as such, re-establishing parasite-based diagnosis becomes a priority.

Treatment

During an outbreak, pre-referral treatment with artesunate suppositories should be provided to severe cases identified in peripheral locations to cover the period of transport to a hospital. Severe cases should be treated with IV artesunate, or if not available, IM artemether. IV Quinine can be used, though its complex administration and need for monitoring (e.g. glycaemia) mean it is not usually a first choice. All severe malaria treatment courses should be followed by a full course of oral ACT (see Chapter 6).

Prevention

- Though treatment is prioritized during malaria outbreaks, preventing new cases is also important (see Chapter 7). The principal interventions to reduce infection risk during emergencies are indoor residual spraying (IRS) where sprayable surfaces are available, and long-lasting insecticidal nets (LLINs). Other insecticide-treated materials, such as insecticide treated plastic sheeting (ITPS), insecticide-treated blankets etc., can be used where LLINs and IRS are not operationally feasible as long as there is local evidence for additional benefit, noting the absence of a general recommendation from WHO.

Vector control – if well planned, targeted and timely – can contribute significantly to reducing infection risk and saving lives. Vector control is most cost-effective when implemented to prevent an epidemic starting or to introduce control in the very early stages of an epidemic. Vector control is logistically demanding and time-consuming. If delayed or mistimed, its effect can be suboptimal (see Annex V).

The most cost-effective interventions for malaria vector control are IRS and LLINs, if used just prior to, or at the start of, an epidemic (i.e. applied well before the epidemic peak and its subsequent natural decline). Programmes must achieve high coverage to affect transmission. IRS requires coverage of at least 80% of dwellings to be fully effective. The peak in malaria cases usually occurs some weeks after vectorial capacity has already peaked and transmission potential has already dropped off. As it usually takes some time to mobilize equipment and spray teams, IRS may not necessarily be the best response in an outbreak. LLINs can provide community vector control in some settings with high population coverage. At lower coverage rates on the other hand, LLINs provide protection only for those sleeping under them and will have limited or no effect in reducing transmission (Magesa et al.; 1991).

Response review

An epidemic response review involves monitoring the evolution of the outbreak and assessing the effectiveness of the response – in terms of how well activities are being carried out, the timing of activities and the level of coverage achieved. Monitoring the roll-out of interventions is an important aspect of outbreak follow-up. The evolution of malaria morbidity and mortality over time can be expressed in terms of proportional incidence and case-fatality rates (see Chapter 4).

If facility-based incidence data is not available, repeat prevalence surveys may be needed whenever there are major changes in the composition of the population (e.g. the arrival of large numbers of non-immune people to endemic areas) and to verify whether the outbreak has been controlled.

References

- Magesa S.M.et al. (1991). Trial of pyrethroid treated bednets in an area of Tanzania holoendemic for malaria. Part 2: Effects on the malaria vector populations. *Acta Tropica*, 49: 97–108.

Finding out more

- WHO Global Malaria Programme website: http://www.who.int/malaria/areas/epidemics_emergencies/en/index.html
- WHO (2004). *Field guide for malaria epidemic assessment and reporting.* Geneva, World Health Organization (WHO/HTM/MAL/2004.1097). Available at http://whqlibdoc.who.int/hq/2004/WHO_HTM_MAL_2004.1097.pdf
- WHO (2004). *Malaria epidemics: forecasting, prevention, early detection and control. From policy to practice. Report of an informal consultation, Leysin, Switzerland, 8–10 December 2003*. Geneva, World Health Organization (WHO/HTM/MAL/2004.1098).
- WHO (2006). *Systems for the early detection of malaria epidemics in Africa: An analysis of current practices and future priorities.* Geneva, World Health Organization. Available at http://whqlibdoc.who.int/publications/2006/9789241594882_eng.pdf
- WHO (2011). *Universal Access to Malaria Diagnostic Testing – an operational manual*. Geneva, World Health Organization. Available at http://whqlibdoc.who.int/publications/2011/9789241502092_eng.pdf
- WHO (2012). *Disease surveillance for malaria control*. Geneva, WHO. Available at http://whqlibdoc.who.int/publications/2012/9789241503341_eng.pdf

CHAPTER 6

Case management

This chapter:

- Describes clinical assessment, including emergency triage for rapid identification and treatment of patients at greatest risk of dying
- discusses confirmatory parasitic diagnosis of malaria using microscopy and rapid diagnostic tests (RDTs)
- describes malaria treatment, including determining treatment choice and management of uncomplicated malaria, treatment failures, and severe malaria with associated complications
- Discusses special groups (e.g. pregnant women, malnourished patients, displaced persons/returnees)

Initial assessment

During humanitarian emergencies in malaria-endemic countries, malaria is a leading cause of mortality among febrile patients, particularly children. If patients with *P. falciparum* malaria do not receive appropriate treatment with an effective antimalarial drug, they may deteriorate and develop severe malaria within a few hours or days. However, it is important to remember that the clinical presentation of both uncomplicated and severe malaria is variable and nonspecific (see Table 6.3), so the differential diagnosis of malaria from other febrile illnesses is very difficult without confirmation of diagnosis by microscopy or rapid diagnostic test (RDT). Additionally, patients often have more than one underlying disease with overlapping clinical presentations (e.g. malaria and pneumonia).

Efforts must be made to improve diagnostic capacity for febrile illnesses by confirmation of the presence of parasites. Reliable RDTs for malaria are now readily available, making diagnostic testing for malaria possible even in emergency settings. If diagnostic testing is not feasible, particularly during the acute stages of an emergency, the most practical approach is to treat febrile patients as suspected malaria cases, with the inevitable consequences of over-treatment of malaria and potentially poor management of other diseases. Knowledge of local disease epidemiology helps in determining the likelihood that any serious febrile illness is due to *P. falciparum*.

Operational aspects of case management

Delivery of care is determined by patient numbers, resources, and access to health services. It is best to use diagnostic tools and medicines that are effective, safe and simple to administer. The essential elements of case management are:

- Triage
- History and physical examination
- Parasitological confirmation of diagnosis
- Antimalarial treatment, with first dose given under observation
- Advise patient or caregiver regarding treatment adherence and when to return
- Management of treatment failures

Emergency triage

Emergency triage is essential. The priority should be to save lives, with the sickest patients treated first. The key is rapid identification, assessment and treatment of emergency danger signs.

Severe malaria is a medical emergency, and patients need immediate referral to a health facility for parenteral treatment and good-quality nursing care. The priority is to identify and strengthen a clinical site capable of managing severe malaria. The risk of developing severe malaria depends on the age and immunity of the patient (see Table 6.1).

Table 6.1 **Groups at high risk of severe malaria and mortality in all transmission areas**

• Non-immune pregnant women
• Young children
• Severely malnourished children and non-immune adults
• HIV-infected persons

Many deaths during the acute phase of emergencies can be avoided with prompt identification and effective treatment of all patients who are severely ill, because of malaria or another condition. However, this is not easy in situations with few skilled staff, limited health care services, and excessive case loads. It is therefore essential to adopt an emergency triage procedure that can effectively identify high-risk patients, and that is simple and quick to perform. The procedures described in the Integrated Management of Childhood Illness (IMCI) guidelines for febrile patients (e.g. Table 6.2) can rapidly identify children who need immediate assessment and treatment (WHO, 2005).

According to the WHO guidelines for the Integrated Management of Adolescent and Adults Acute Care (2009), all patients should receive an initial Quick Check for Emergency Signs. Any patients who present one or more of the following conditions require rapid referral after initial emergency treatment:

- Airway and breathing: Appears obstructed or central cyanosis (blue mucosa) or severe respiratory distress;
- Circulation: Weak and fast pulse or capillary refill longer than 2 seconds;
- Unconscious/repeated convulsing: Convulsing (now or recently), or unconscious;
- Chest pain, severe abdominal pain, neck pain or severe headache;
- Any fever with one or more of the following:
  - confusion, agitation, lethargy
  - stiff neck
  - very weak (not able to stand or to walk unaided)
  - not able to drink
  - severe abdominal pain
  - fast and deep breathing or severe respiratory distress.

Table 6.3 lists clinical features associated with malaria. Although none of the clinical signs listed is unique to malaria, those related to severe malaria are a clear indication that a patient needs immediate investigation and treatment (see Table 6.4 and Managing severe *P. falciparum* malaria below).

Confirmatory malaria diagnosis

Purpose of confirmatory diagnosis

Diagnosing malaria on the basis of clinical features alone can be highly inaccurate and is likely to result in significant over-treatment. It is therefore important to establish, at the start of any emergency intervention, capacity for confirmatory diagnostic testing. Effective confirmatory diagnostic testing can help to:

- identify patients who need antimalarial treatment;
- reduce unnecessary use of antimalarial drugs for patients without malaria;

Table 6.2 **IMCI guidelines for the management of febrile children**

Does the child have fever?
(by history or feels hot or temperature 37.5 °C[a] or above)

If yes, decided malaria risk: high or low
Then ask:
- For how long?
- If more than 7 days, has fever been present every day?
- Has the child had measles within the last 3 months?

Look and feel:
- Look or feel for stiff neck
- Look for runny nose

Look for signs of measles:
- Generalised rash, and
- one of these: cough, runny nose, or red eyes

Look for any other cause of fever
Do a malaria test if NO general danger sign or stiff neck
- HIGH MALARIA RISK: do a malaria test in all fever cases
- LOW MALARIA RISK: do a malaria test if no obvious cause of fever
- Look for mouth ulcers

CLASSIFY FEVER

Malaria risk	Signs	Classify	Treatment
HIGH MALARIA RISK	• Any general danger sign or • Stiff neck	VERY SEVERE FEBRILE DISEASE	• Give first dose of quinine or artesunate for severe malaria • Give first dose of an appropriate antibiotic • Treat the child to prevent low blood sugar • Give one dose of paracetamol in clinic for high fever (≥38.5 °C) • Refer URGENTLY to hospital
	• Malaria test POSITIVEb	MALARIA	• Give recommended first-line oral antimalarial • Give one dose of paracetamol in clinic for high fever (≥38.5 °C) • Advise mother when to return immediately • Follow-up in 3 days if fever persists • If fever is present every day for more than 7 days, refer for assessment
	• Malaria test NEGATIVE • Runny nose PRESENT, or • Measles PRESENT, or • Other cause of fever PRESENT	FEVER: NO MALARIA	• Assess for possible bacterial cause of fever[c] and treat with appropriate drugs • Give one dose of paracetamol in clinic for high fever (≥38.5 °C) • Advise mother when to return immediately • Follow-up in 2 days if fever persists • If fever is present every day for more than 7 days, refer for assessment
LOW MALARIA RISK	• Any general danger sign or • Stiff neck	VERY SEVERE FEBRILE DISEASE	• Give first dose of quinine or artesunate for severe malaria • Give first dose of an appropriate antibiotic • Treat the child to prevent low blood sugar • Give one dose of paracetamol in clinic for high fever (≥38.5 °C) • Refer URGENTLY to hospital
	• Malaria test POSITIVEb	MALARIA	• Give recommended first-line oral antimalarial • Give one dose of paracetamol in clinic for high fever (≥38.5 °C) • Advise mother when to return immediately • Follow-up in 3 days if fever persists • If fever is present every day for more than 7 days, refer for assessment
	• Malaria test NEGATIVE • Runny nose PRESENT, or • Measles PRESENT, or • Other cause of fever PRESENT	FEVER: NO MALARIA	• Assess for possible bacterial cause of fever[c] and treat with appropriate drugs • Give one dose of paracetamol in clinic for high fever (≥38.5 °C) • Advise mother when to return immediately • Follow-up in 2 days if fever persists • If fever is present every day for more than 7 days, refer for assessment
NO MALARIA RISK and no travel to malaria risk area	• Any general danger sign or • Stiff neck	VERY SEVERE FEBRILE DISEASE	• Give first dose of an appropriate antibiotic • Treat the child to prevent low blood sugar • Give one dose of paracetamol in clinic for high fever (≥38.5 °C) • Refer URGENTLY to hospital
	• No general danger sign • No stiff neck	FEVER	• Assess for possible bacterial cause of fever[c] and treat with appropriate drugs • Give one dose of paracetamol in clinic for high fever (≥38.5 °C) • Advise mother when to return immediately • Follow-up in 2 days if fever persists • If fever is present every day for more than 7 days, refer for assessment

[a] These tempertures are based on axillary temperature. Rectal temperature readings are approximately 0.5 °C higher.
[b] If no malaria test available: high malaria risk – classify as malaria; low malaria risk and no other obvious cause of fever – classify as malaria.
[c] Look for local tenderness, refusal to use a limb, hot tender swelling, red tender skin or boils, lower abdominal pain or pain in passing urine.

Source: Integrated Management of Childhood Illness (IMCI)

Table 6.3 **Clinical features of uncomplicated and severe malaria**

Uncomplicated malaria	Severe malaria
Clinical features may include: • Fever • Headache • Vomiting • Diarrhoea • Cough • Influenza-like symptoms (e.g. chills, muscle pains) • Febrile convulsions (in children)	*Clinical features may include:* • Prostration (i.e. generalized weakness so the patient is unable to sit or to walk) • Impaired consciousness or unarousable coma, not attributable to another cause • Multiple convulsions (>2 in last 24 hours) • Severe normocytic anaemia • Hypoglycaemia • Metabolic acidosis • Deep breathing, respiratory distress • Acute renal injury • Acute pulmonary oedema and adult respiratory distress syndrome (ARDS) • Circulatory collapse or shock, • Abnormal bleeding (e.g. bruising, bleeding gums, haemoglobinuria) • Jaundice plus evidence of vital organ dysfunction • Hyperlactataemia

Important: These severe manifestations can occur singly or, more commonly, in combination in the same patient.

— identify patients who need management for other illnesses;
— improve malaria case detection and reporting;
— provide confirmation of treatment failures.

Parasitological confirmation of suspected malaria before treatment is now recommended for all patients. There are currently two widely available options for confirmatory diagnostic testing of malaria:

— light microscopy
— rapid diagnostic tests (RDTs).

The characteristics and potential uses of these two methods are outlined below.

An interactive guide, allowing users to identify RDTs that have been evaluated by the WHO Product Testing Programme based on different diagnostic performance parameters, is available online from the Foundation for Innovative New Diagnostics (FIND) at: http://www.finddiagnostics.org/programs/malaria/find_activities/rdt_quality_control/product_testing/malaria-rdt-product-testing/index.jsp.

Table 6.4 **Comparison of microscopy and RDTs**

	Microscopy	RDTs
Requirements		
Equipment	Microscope	None
Electricity	Preferred, not necessary	None
Supplies	Lancets, slides, alcohol swabs, staining reagents and supplies, water	Lancets, blood collecting devices, alcohol swabs (included in some kits)
Transport/storage conditions	Reagents stored out of direct sunlight	Avoid exposure to high temperature (4–30 °C recommended)
Quality assurance	Periodic re-reading of percentage of slides by expert microscopist and supervision of microscopists	Lot testing of kits, monitoring of storage temperature, supervision of health workers
Performance		
Test duration	Usual minimum 60 minutes	15–20 minutes
Labour-intensiveness	High	Low
Dependence on individual competence	High	Low
Direct costs		
Cost per test	US$0.12–0.40	US$0.60–1.00
Technical specifications		
Detection threshold	5–10 parasites/μL	40–100 parasites/μL
Detection of all 4 species	Yes	Some RDTs
Quantification	Possible	Not possible
Differentiation between Pv, Po, and Pm	Possible	Few can differentiate Pv from others
Differentiation between sexual and asexual stages	Possible	Not possible
Detection of Pf sequestered parasites	No	Possible
Antigen persistence	No	Yes (kits detecting HRP2 antigen)

Table 6.4 **Continued**

Advantages and disadvantages of microscopy or RDTs in special situations		
Routine diagnosis and patient management	Determination of parasite density for management of severe malaria	More easily implemented in emergency situations
Investigating suspected treatment failure	Recommended	Not recommended, as some HRP2 kits can remain positive several weeks after parasite clearance
Drug efficacy studies	Recommended	Not recommended
High case loads	1 skilled microscopist can prepare/read 40–60 slides in a day	1 trained health worker can process up to 100 RDTs in a day
Community surveys or rapid assessments	May be time consuming and labour intensive	Suitable for screening large numbers of people in the field

Source: Updated and adapted from WHO (2011b)

Light microscopy

Microscopy has long been considered the "gold standard" for malaria diagnosis. When performed by a skilled technician, microscopy is more sensitive than RDTs for the detection of low levels of *P. falciparum* parasitaemia (<100 parasites/µl) and can differentiate between the various malaria species. However, when poorly performed, microscopy produces unreliable results with lower sensitivity and specificity than RDTs. Effective microscopy for malaria parasites requires: (i) good-quality equipment and reagents; (ii) skilled technicians who can prepare and stain films, identify parasites, and differentiate between Plasmodium species; and (iii) rigorous technical supervision and quality control. However, since these may not be available in emergencies, it is often necessary for emergency partners to establish capacity themselves (see WHO 2011b).

Rapid diagnostic tests

In the acute emergency phase, limited time and resources make RDTs preferable to microscopy for confirmation of clinical diagnosis in low-transmission areas, and for confirming malaria in severely ill patients in moderate to high-transmission areas. Malaria RDT selection should be based on the prevalence of malaria species in the country (WHO 2011). Three geographical zones have been defined:

- Zone 1, in which *P. falciparum* is predominant and the majority of non-falciparum species cause mixed infection with *P. falciparum* (e.g. most areas of sub-Saharan Africa and lowland Papua New Guinea);
- Zone 2, where *P. falciparum* and non-falciparum infections occur commonly as single-species infections (e.g. most endemic areas of Asia and the Americas and isolated areas in the Horn of Africa); and
- Zone 3, where only non-falciparum infections occur (e.g. mainly *P. vivax*-only areas of East Asia, central Asia, South America, and some highland areas elsewhere).

In Zone 1, RDTs that detect only *P. falciparum* are generally preferable. In Zone 2, RDTs that detect all species and distinguish *P. falciparum* from non-falciparum infections are necessary. In Zone 3, RDTs that detect non-falciparum species alone are appropriate (i.e. pan-specific or *P. vivax*-specific RDTs).

The type of antigen targeted depends on the species to be detected (see Table 6.5). To detect *P. falciparum*, tests targeting HRP2 are generally preferable, as they are more sensitive than those that detect pLDH. To detect non-falciparum species, tests targeting Aldolase, pLDH specific to non-falciparum species, or pLDH common to all species, are recommended.

Ideally, commercially available RDTs should be selected in accordance with national malaria control programme recommendations and, selection should be based on the results of the WHO product testing programme (WHO 2012).

Table 6.5 **Antigen targets of rapid diagnostic tests (RDTs)**

***Plasmodium* species**	**HRP2**	**pLDH**				
		pLDH–Pf	**pLDH–pan**	**pLDH–Pvom**	**pLDH–Pv**	**Aldolase**
P. falciparum	x	x	x			x
P. vivax			x	x	x	x
P. malariae			x	x		x
P. ovale			x	x		x

NB: HRP2 – histidine rich protein 2; pLDH – *Plasmodium* lactate dehydrogenase; Pf – *P. falciparum*; pan – all *Plasmodium* species; Pvom – *P. vivax, P. ovale,* and *P. malariae*; Pv – *P. vivax*
Source: WHO (2011)

Malaria treatment

Wherever possible, a confirmatory parasitological diagnosis of suspected malaria cases should be conducted before initiating malaria treatment. The signs and symptoms of malaria are varied. Virtually all non-immune individuals will experience fever. Other frequent symptoms include chills, sweating, headache, anorexia, myalgia, fatigue, nausea, abdominal pain, vomiting, and diarrhoea. Anaemia, thrombocytopenia, splenomegaly, hepatomegaly, and jaundice can develop. Case definitions for malaria can be found in Table 4.2:

- *Suspected malaria* – a patient with a fever, or history of fever in the last 48 hours, who is currently residing in or has come from a malaria-endemic area;
- *Uncomplicated malaria* – a patient with a fever, or history of fever in the last 48 hours, with a positive confirmatory parasitological test and no signs of severity or evidence of vital organ dysfunction;
- *Severe malaria* – a patient with a positive confirmatory parasitological test and one or more symptoms of severe disease (see Table 6.3).

Choosing antimalarial drugs

Wherever possible, national policy and malaria treatment guidelines should be followed. In emergencies, limited access to health care and poor security may mean that a health worker sees a patient only once. Since patient adherence to multi-day therapies may be poor and can lead to treatment failure, it is vital that treatment regimens are as simple and convenient as possible and clearly and accurately explained to each patient. The choice of drugs used in malaria treatment depends on:

- local parasite species;
- antimalarial susceptibility of local parasite strains;
- drug therapeutic efficacy;
- drug safety (i.e. few side effects);
- safety for special groups such as young children and pregnant women;
- acceptability to both providers and consumers (e.g. packaging, taste, specific child-friendly formulations);
- simplicity of dosage regimes to encourage treatment adherence;
- availability on the market with a minimal lead time;
- authorization for use by national health authorities;
- affordability for consumers (in cases where medicines are not being provided free-of-charge).

During humanitarian emergencies, populations are more vulnerable to malaria, and the first-line treatment for falciparum malaria should be of the highest possible efficacy. Artemisinin-based combination therapies (ACTs) are currently the best choice for treating uncomplicated falciparum malaria. The efficacy of an individual ACT depends on resistance to the artemisinin partner drug. Any ACT chosen for deployment in a humanitarian emergency should have an efficacy of at least 95%. ACT options currently recommended by WHO, in alphabetical order, are:

- Artemether + Lumefantrine (AM/LM)
- Artesunate + Amodiaquine (AS+AQ)
- Artesunate + Mefloquine (MAS)
- Artesunate + Sulfadoxine/Pyrimethamine (AS+SP)
- Dihydroartemisinin + Piperaquine (DP)

Fixed-dose/coformulated drug combinations should be used where possible, particularly in humanitarian emergencies, since their use improves patient compliance as compared to co-blistered combinations. The most up-to-date list of recommended ACTs can be found in the WHO Guidelines for the Treatment of Malaria posted on the Global Malaria Programme.

WHO has established a system for pre-qualification of antimalarial medicines including ACTs, and provides guidelines for pharmacovigilance. Medicines with the highest efficacy and safety record should always be deployed, especially in situations where the population is stressed, suffers from malnutrition, and may have low levels of immunity. WHO treatment guidelines list recommended pharmaceutical products selected on the basis of clinical efficacy, safety, and suitability for large-scale use; medicines for use during humanitarian emergencies should always be chosen from this list.

Correct transport and storage of drugs, particularly ACTs, is extremely important, as is abiding by expiry dates to avoid degradation of active pharmaceutical ingredients,.

If no information on local drug efficacy is available, an ACT should be chosen from those that have the highest cure rates (i.e. over 95% efficacious in most settings). In chronic emergencies, every other year therapeutic efficacy testing of the chosen ACT should be conducted in sentinel sites to determine that it remains above 95% efficacious. Information on antimalarial drug resistance and testing protocol is available at http://www.who.int/malaria/diagnosis_treatment/resistance/en/index.html.

A coordinated approach to procurement and distribution is essential if adequate supplies of good-quality antimalarial drugs are to be provided regularly. Standardized purchase and supply of recommended, good-quality antimalarial drugs are the responsibility of all emergency partners.

Managing uncomplicated falciparum malaria

In high transmission zones, all cases of fever suspected as malaria should have a confirmatory parasitological diagnostic test before treatment is prescribed. Test results should always be considered accurate, and if negative, another cause of fever should be urgently sought in order that correct treatment can be administered. It is important to remember that malaria may be accompanied by other serious conditions (e.g. pneumonia) and a positive malaria test does not exclude bacterial infections that may also require antibiotic treatment. If the patient does not have severe illness, the test is negative, and no other cause can be found, the patient should be asked to return within two days – or sooner if their condition deteriorates.

In low and moderate transmission zones, a careful examination of the patient should be conducted to exclude other causes of fever before parasitological testing is initiated. In areas with both *P. falciparum* and *P. vivax*, microscopy or combination RDTs should be used to ensure that the correct treatment is given.

ACTs are the treatment of choice for uncomplicated *P. falciparum* malaria and patients with mixed infections. ACTs are indicated for treatment of uncomplicated falciparum malaria during the first trimester of pregnancy if this is the only treatment immediately available, or if there is uncertainty about compliance with a 7-day treatment (e.g. quinine plus clindamycin). Treatment of uncomplicated falciparum malaria during the second and third trimesters of pregnancy should be with one of the following:

- ACTs known to be effective in the country/region;
- artesunate plus clindamycin for seven days; or
- quinine plus clindamycin for seven days.

See Annex VI for treatment regimens.

Artemisinin derivatives are contraindicated in cases of a proven allergy. Oral monotherapies (e.g. chloroquine, amodiaquine, sulfadoxine/pyrimethamine, quinine, and especially artemisinin or its derivatives) should never be used to treat uncomplicated falciparum malaria or mixed falciparum infections.

Supportive and ancillary care for uncomplicated malaria

When an antimalarial drug is given to a young child, the chance of the child vomiting the medication will be lower if fever (i.e. axillary temperature above 37.5 °C) is first treated with an antipyretic and tepid sponging. An antipyretic such as paracetamol should be used. Aspirin is not recommended for use in children because of the increased risk of Reye's syndrome. Mothers should be instructed on how to administer medication successfully, especially to young infants.

Children should remain at the clinic for an hour after the first dose of antimalarial drug is administered (see Annex IV for information concerning the safety of antimalarial drugs in young children). If it is not possible for a child to wait at the clinic, the parent or carer should be told to return if the child vomits within the first hour after taking the medication. If a child vomits within 30 minutes of taking the first dose, administer a full replacement dose. If vomiting occurs between 30 minutes and 1 hour, give a half dose. A child who vomits the drug more than once should be managed as a severely ill patient, as "vomiting everything" is a danger sign. It is important to watch for signs of dehydration and hypoglycaemia, especially in very young or malnourished children and pregnant women, and to give appropriate treatment if necessary. Caregivers should be told that it is important for infants to breastfeed frequently, and for older children to drink plenty of fluids, to prevent dehydration.

Carers and/or patients should be instructed to return to the health facility for immediate care if danger signs appear (see Table 6.2) or symptoms worsen. Carers and/or patients should also be instructed to return to the health facility if symptoms have not resolved after completion of treatment.

Managing treatment failures

Recurrence of *P. falciparum* malaria can result from either re-infection or recrudescence/failure. Treatment failure is defined as the failure to effectively clear malaria parasites from the blood or resolve clinical symptoms despite administration of an antimalarial medicine. Treatment failure may result from a number of causes.

- *Vomiting or poor absorption* – the patient or carer may have been confused about what to do when treatment was vomited.
- *Poor prescribing practice* – prescription or sale of an incomplete course of antimalarial treatment increases the risk of treatment failure.
- *Poor adherence* – this may be caused by several factors, including:
  - information on the correct regimen may have been unclear to the patient or carer;
  - patients may choose to stop taking the treatment once initial symptoms have resolved (particularly problematic with drugs that pro-

duce very rapid symptoms clearance), or to share the remaining treatment with family/friends;
  - in areas with cost-recovery systems, many patients cannot afford to buy a complete course of treatment;
  - where drugs are not co-formulated, patients may discard one of the medicines of the combination that is associated with side effects.
- *Drug quality* – the use of antimalarial drugs that do not contain the recommended amount of active ingredient is a particular risk if patients are receiving treatment from the informal private sector. ACT quality should be ensured by: (i) sourcing from reputable pharmaceutical companies and procurement agencies; (ii) proper storage at all levels of the distribution system to avoid degradation due to exposure to high temperatures; and (iii) not administering drugs beyond their expiry dates.
- *Drug resistance* – is the ability of a parasite to survive and/or multiply despite the administration and absorption of a drug given in doses equal to or higher than those usually recommended, but within the tolerance of the subject.

Treatment failure within 14 days of receiving an ACT is very unusual. A careful history and examination is necessary to ascertain any non-malarious causes of symptoms. For example, was the complete course of ACT taken, accompanied by food where appropriate, and with no vomiting in the first hour following any dose? All treatment failures should be confirmed parasitologically, preferably by blood slide examination. It is important to keep a record of all true treatment failures and inform health authorities, as these failures need to be confirmed by drug efficacy studies that may lead to changes in first-line antimalarial treatment.

- Treatment failures after 14 days of initial treatment, should be considered re-infections in the absence of PCR genotyping and another dose of ACT can be given (avoid reuse of mefloquine within 60 days, due to increased risk of neuropsychiatric reactions: in case AS+MQ was the initial treatment an alternative ACT not containing mefloquine should be given).
- Treatment failures within 14 days of initial treatment should be treated with a second-line antimalarial, in order of preference:
  - an alternative ACT known to be effective in the region;
  - artesunate plus tetracycline *or* doxycycline *or* clindamycin (for a total of 7 days);
  - quinine plus tetracycline or doxycycline or clindamycin (for a total of 7 days).

Doxycycline and tetracycline, but not clindamycin, are contraindicated for pregnant women and children under 8 years old. Drug choice will depend on the antimalarial drug already taken, availability of alternatives, contraindications, and operational feasibility. Adherence to the full course of re-treatment is vital, and is improved if detailed instructions are given to patients, carers and health personnel.

Managing severe falciparum malaria

The main treatment objective for severe malaria is to prevent the patient from dying. Secondary objectives are prevention of disabilities and recrudescence. In humanitarian emergencies, where patient numbers are high and there are many late presentations, effective triage with immediate resuscitation and treatment are essential (Table 6.6). The clinical features of severe falciparum malaria are provided in Table 6.3.

Table 6.6 **Supportive treatment for patients with severe malaria**

- Clear the airway and check that the patient is breathing;
- Establish intravenous (IV) access;
- Treat convulsions lasting 5 minutes or more, (see below);
- Take blood for malaria parasites, blood glucose and haemoglobin (urea and electrolytes, blood gas and blood culture are extremely useful, but unlikely to be feasible in most humanitarian emergencies);
- Treat hypoglycaemia[a] (blood glucose <2.2 mmol/l), (see Table 6.7);
- Rapidly assess circulation, hydration and nutritional status, and resuscitate as necessary with normal (0.9%) saline, (see below);
- In children, if haemoglobin is <4 g/dl, transfuse blood;
- In children, if haemoglobin is 4–6 g/dl, transfuse blood if the patient also has respiratory distress, impaired consciousness, hyperparasitaemia (>20% parasitized RBCs), shock, or heart failure;
- In children in low transmission settings and in adults, if haemoglobin is <7 g/dl, transfuse blood;
- For unconscious patients, insert a nasogastric tube and aspirate stomach contents to prevent aspiration pneumonia;
- Place the patient in the lateral or semi-prone position, and perform a lumbar puncture to exclude meningitis;
- Start antimalarial drug treatment, (see below);
- Start antibiotic therapy, (see below).

[a] Threshold for correction of hypoglycaemia in children is blood glucose < 3 mmol/l.

Pre-referral treatment – For emergency pre-referral treatment of severe malaria, or of a patient who cannot tolerate oral medication, artesunate can be administered rectally before transport to a facility where parenteral treatment of severe malaria can be provided. Once the patient is able to tolerate oral medication, treatment must be completed with a full course of ACT.

Antimalarial drug treatment – Parenteral artesunate is the treatment of choice for severe malaria, given either by the intravenous or intramuscular route. Artemether and quinine are acceptable alternatives if injectable artesunate is not readily available. Parenteral treatment should be given for at least 24 hours, or until the patient can tolerate oral medications, and should always be followed by a full course of ACT (see Annex VI for detailed drug regimens).

Adjunctive treatment of complications associated with severe malaria

Coma, convulsions, hypoglycaemia (blood glucose <2.2 mmol/l), severe anaemia, shock, pulmonary oedema, and acute renal failure are common complications of severe malaria that require immediate management (Table 6.7).

Blood transfusion – It is *essential* to ensure a safe supply of blood for transfusion. Blood should be cross-matched and screened for HIV, malaria and hepatitis B. Local laboratory facilities must therefore be able to perform compatibility testing (i.e. cross-matching) and screening for HIV, malaria and, if possible, hepatitis B. If a safe supply cannot be assured, transfusion should be restricted to patients with severe anaemia and signs of acute failure (shock, respiratory distress). If suitable donors without malaria infection cannot be found, blood should be administered with an antimalarial treatment.

Concomitant antibiotics – There is considerable clinical overlap between septicaemia, pneumonia and severe malaria and these conditions may coexist. *In children* with suspected severe malaria with associated alterations in the level of consciousness, broad spectrum antibiotic treatment should therefore be started immediately along with antimalarial treatment, and should be completed unless a bacterial infection is excluded. *In adults* with severe malaria, antibiotics are recommended if there is evidence suggestive of bacterial co-infection (e.g. shock, pneumonia). *All patients* with clinical evidence of bacterial infection (e.g. pneumonia, dysentery) should receive antibiotic therapy according to local treatment protocols.

Fluid management – Fluid requirements should be assessed carefully and individually to avoid over and under-hydration. In adults, both dehy-

Table 6.7 **Managing complications of severe malaria**

Manifestation/complication	Immediate management[a]
Coma (cerebral malaria)	Maintain airway, place patient on his side or her side, exclude other treatable causes of coma (e.g. hypoglycaemia, bacterial meningitis); avoid harmful ancillary treatment, such as corticosteroids, heparin and adrenaline; intubate if necessary.
Hyperpyrexia	Administer tepid sponging, fanning, a cooling blanket and antipyretic drugs. Paracetamol is preferred over more nephrotoxic drugs (e.g. NSAIDs[b]).
Convulsions	Maintain airways; treat promptly with intravenous or rectal diazepam or intramuscular paraldehyde. Check blood glucose.
Hypoglycaemia	Check blood glucose, correct hypoglycaemia and maintain with glucose-containing infusion.
Severe anaemia	Transfuse with screened fresh whole blood.
Acute pulmonary oedema[c]	Prop patient up an an angle of 45°, give oxygen, give diuretic, stop intravenous fluids, intubate and add positive end-expiratory pressure/continuous positive airway pressure in life threatening hypoxaemia.
Acute renal failure	Exclude pre-renal causes, check fluid balance and urinary sodium; if in established renal failure add haemofiltration or haemodialysis , or if unavailable, peritoneal dialysis.
Spontaneous bleeding and coagulopathy	Transfuse with screened fresh whole blood (cryoprecipitate, fresh frozen plasma and platelets, if available); give vitamin K injection.
Metabolic acidosis	Exclude or treat hypoglycaemia, hypovolaemia and septicaemia. If severe add haemofiltration or haemodialysis
Shock	Suspect septicaemia, take blood for cultures; give parenteral broad-spectrum antimicrobials, correct haemodynamic disturbances.

[a] It is assumed that appropriate antimalarial treatment will have been started in all cases
[b] Non-steroidal anti-inflammatory drugs
[c] Prevent by avoiding excess hydration
Source: WHO (2010)

dration and rehydration should be managed cautiously as relatively small changes in fluid levels can mean the difference between over-hydration and increased pulmonary oedema risk, and under-hydration, contributing to shock, worsening acidosis and renal impairment. Careful and frequent evaluations of jugular venous pressure, peripheral perfusion, venous filling, skin turgor and urine output should be made.

Use IV fluids containing 0.9% (normal) saline and 5–10 % dextrose, changing to 10% dextrose if the patient becomes hypoglycaemic. In most humanitarian emergencies, with large numbers of patients and limited numbers of nursing staff, close observation of patients may be impossible and it is safest to give maintenance fluids by nasogastric tube. The carer can help to administer 4-hourly nasogastric feeds of milk or diluted porridge, making it possible to provide calories in addition to fluid – an advantage as many patients are likely to be malnourished. However, inhalation pneumonia must be prevented. Rapid fluid boluses are contraindicated in severe malaria resuscitation.

Anaemia – Anaemia is commonly associated with malaria and is particularly serious for young children and pregnant women. To assess patients of all ages, check for palmar and conjunctival pallor. Time constraints are likely to necessitate starting oral treatment on the basis of clinical signs alone. Where possible, measure blood haemoglobin (Hb) level – HemoCue® can be used in most settings. Hb <5 g/dl indicates severe anaemia, requiring urgent referral to an inpatient facility for further assessment and possible blood transfusion. Hb of 5–10 g/dl requires oral treatment.

Treating anaemia

Anaemia is a condition in which haemoglobin production or the number of red blood cells is diminished in relation to the age, gender, residential elevation above sea level, smoking behaviour, and stage of pregnancy. WHO recommends that anaemia be diagnosed by measuring haemoglobin concentrations and monitored until it disappears. Clinical symptoms such as weakness and pallor of the eyelids, tongue, nail beds, or palms appear in severe cases. It is acknowledged that between 50 and 60% of anaemia cases are due to iron deficiency. To overcome other nutritional causes of anaemia, vitamins and minerals may be added to iron and folic acid supplements without exceeding the daily recommended intake dose (WHO, 2004).

The doses below are recommended by WHO for the treatment of anaemia. However, intermittent iron supplementation is recognized as a public health intervention that can be used to prevent anaemia in children, menstruating women, and non-anaemic pregnant women (WHO, 2011). In malaria-endemic areas, the provision of iron supplements should be implemented in conjunction with adequate measures to prevent, diagnose and treat malaria.

In children

- Give 3 mg elemental iron/kg/day in the form of drops, chewable tablets or powders to infants and younger children (approximately 30 mg/day at 24 months of age). For children 60 months of age and older give 60 mg elemental iron day in the form of tablets, pills or powders. 1 mg of elemental iron equals 3 mg of ferrous fumarate, 5 mg of ferrous sulfate heptahydrate or 8.4 mg of ferrous gluconate.
- Give 250 μg (0.25 mg) of folic acid to children 60 months of age and older.
- Treat all children over 2 years of age presumptively for intestinal worms with a single 500-mg dose of mebendazole or 400-mg dose of albendazole (unless either of these has been given in the previous 6 months); Advise caregivers about good feeding practices. Meat, poultry, fish or eggs should be eaten daily, or as often as possible, because they are rich sources of many key nutrients such as iron and zinc. Diets that do not contain animal source foods (meat, poultry, fish or eggs, plus milk products) cannot meet all nutrient needs at this age unless fortified products or nutrient supplements or powders are used. During humanitarian emergencies, fortified products or nutrient/powder supplements may be more practical than providing animal source foods.

In non-pregnant adults

- Give 60 mg elemental iron twice a day in two separate doses (i.e morning and evening, for a total dose of 120 mg elemental iron). 60 mg of elemental iron equals 300 mg of ferrous sulfate heptahydrate, 180 mg of ferrous fumarate or 500 mg of ferrous gluconate;
- Give 400 μg (0.4 mg) folic acid per day;
- Treat presumptively for intestinal worms with a single 500mg dose of mebendazole or 400 mg dose of albendazole. Diagnosis should be confirmed by microscopy, if time and facilities allow.

In pregnant women

- Give 120 mg elemental iron daily in two separate doses of 60 mg (i.e morning and evening). When anaemia is no longer present reduce the dose to 60 mg elemental iron per day. 60 mg of elemental iron equals 300 mg of ferrous sulfate heptahydrate, 180 mg of ferrous fumarate or 500 mg of ferrous gluconate.

Supportive and ancillary care for patients with severe malaria

Regular observation of patients with severe malaria is critical, because the clinical situation may change quickly. The most important observations are pulse, respiratory rate and pattern, blood pressure, temperature, and level of consciousness. If there is any deterioration in consciousness, it is essential to check for hypoglycaemia and for a significant fall in haemoglobin, because these are amenable to treatment.

In all patients:

- Check Hb and parasitaemia daily for the first 3 days and before discharge from hospital.

In unconscious patients:

- Ensure the airway is clear;
- Nurse the patient in the lateral or semi-prone position to avoid aspiration of fluid;
- Insert a nasogastric tube and aspirate the stomach contents into a syringe every 4 hours, to reduce the risk of aspiration pneumonia;
- Monitor temperature, pulse, respiration, blood pressure, blood glucose, and level of consciousness at least every 4 hours until the patient is out of danger;
- Suspect raised intracranial pressure in patients with irregular respiration, abnormal posturing, worsening coma, unequal or dilated pupils, elevated blood pressure and falling heart rate, or papilloedema. In all such cases, nurse the patient in a supine posture with the head of the bed raised;
- Check blood glucose and Hb if the level of consciousness deteriorates;
- Assess fluid balance daily if possible (daily weight gives a rough indication of overall fluid balance);
- Report immediately any deterioration in level of consciousness, occurrence of convulsions or changes in behaviour;
- Turn unconscious patients every 2 hours to prevent pressure sores;
- If rectal temperature rises above 39 °C, remove patient's clothes, give oral or rectal paracetamol, and use tepid sponging and fanning to reduce temperature;
- Note appearance of red or black urine (haemoglobinuria), and assess haematocrit, as severe anaemia may develop rapidly.

Managing mixed Plasmodium infections

Mixed malaria infections should be treated with an appropriate ACT. The addition of 14 days of primaquine is indicated in the case of *P. vivax* or *P. ovale* mixed infections (see *Managing relapsing malaria infections* below).

Managing non-falciparum malaria

Chloroquine remains the treatment of choice for non-falciparum malaria in most settings. With the exception of AS+SP, which is not highly effective against *P. vivax*, ACTs cure all types of malaria. However, chloroquine is still effective and considerably cheaper for treating most cases of *P. vivax*, *P. malariae* and *P. ovale*. In chloroquine-sensitive areas, a 3-day course of treatment kills most stages of all non-falciparum species. In specific geographical areas (particularly on the island of Borneo) the monkey parasite *P. knowlesi* can cause malaria in humans. Chloroquine is fully effective, but treatment should be started immediately, as uncomplicated *P. knowlesi* cases can rapidly lead to severe disease with a high case fatality rate.

Evidence of chloroquine efficacy in managing vivax malaria should be sought for the specific area of deployment. Chloroquine-resistant *P. vivax* has been confirmed in Brazil, Bolivia, Ethiopia, Indonesia, Malaysia (Borneo), Myanmar, Papua New Guinea, Peru, the Solomon Islands, and Thailand. Chloroquine-resistant *P. malariae* has been reported in Indonesia.

Managing relapsing malaria infections

Both chloroquine and ACTs only kill parasites in red blood cells, and not pre-erythrocyte forms and the hypnozoites of *P. vivax* and *P. ovale* in the liver that are responsible for relapses. At present, primaquine is the only drug available to eliminate hypnozoites and prevent relapses of non-falciparum malaria. Radical treatment of relapsing malaria (i.e. *P. vivax*, *P. ovale*) requires 14 days of primaquine treatment. Primaquine may be given concurrently with an active blood schizonticide, such as chloroquine, from the first day of treatment.

Anti-relapse treatment is unnecessary for patients living in endemic areas with sustained high transmission. In such cases, relapses cannot be distinguished from reinfections and such patients should be treated with an effective blood schizonticide for each symptomatic recurrence of malaria.

Primaquine anti-relapse treatment is usually not feasible during the acute emergency phase, but may be appropriate in chronic emergencies or during resettlement and repatriation. Adherence to the required 14-day treatment course may prove challenging but shorter anti-relapse treatment courses (e.g. the 5-day course adopted in some Asian countries) are ineffective and are not recommended.

Primaquine is contraindicated for pregnant women, children under 12 months old, or patients with glucose-6-phosphate dehydrogenase (G6PD) deficiency. Daily administration of primaquine for a 14-day radical cure of vivax malaria should not be given to any patient with G6PD deficiency, because of the risk of intravascular haemolysis. Fetuses are G6PD deficient and prevalence of G6PD deficiency is also high in Asia and some parts of Africa. Patients should be tested for G6PD deficiency before treatment with primaquine. Current G6PD testing kits are relatively inexpensive and simple to use, but are laboratory based. G6PD field tests are under development.

Managing malaria in special groups

Pregnant women, children under 5 years of age, severely malnourished people, people living with HIV (PLHIV) and displaced populations generally require additional attention.

Pregnant women

Plasmodium falciparum malaria is a major cause of maternal, perinatal and newborn morbidity and mortality. The clinical effects of falciparum malaria depend on the immune status of women, which is determined by previous exposure to malaria and parity. *P. vivax* infection can cause negative health outcomes for both mother and fetus, particularly maternal anaemia and low birth-weight. To address malaria risk during pregnancy, WHO recommends LLINs, intermittent preventive treatment in pregnancy (IPTp – in areas of sub-Saharan Africa with moderate-to-high transmission of malaria), and effective case management of clinical malaria.

In low falciparum transmission settings, pregnant women have little pre-existing immunity and malaria usually presents as an acute illness with detectable peripheral parasitaemia. Uncomplicated malaria can progress quickly to severe malaria, with high case fatality rates and risk of death for the fetus (i.e. abortion or stillbirth). Pregnant women without immunity to *P. falciparum* are 2–3 times more likely to develop severe disease, and approximately three times more likely to die than non-pregnant women. In these settings, management of malaria risk in pregnancy should be based on prompt diagnostic testing and effective treatment.

In settings with moderate-to-high transmission of *P. falciparum*, pregnant women have a degree of pre-existing immunity and may develop malaria-related anaemia with no microscopically-detectable parasites in the peripheral blood, although the placenta may be heavily infected. The main effect of placental infection on the fetus is low birth-weight and increased risk of neonatal mortality. These effects are observed frequently in HIV-negative women in their first and second pregnancies. Among pregnant PLHIV, malaria infections increase maternal anaemia risk and low birth-weight risk in all pregnancies. In these settings, management of malaria risk in pregnancy should be based on prevention using LLINs and prompt diagnosis and treatment of anaemia and uncomplicated malaria. Saving the mother should take priority when treating severe malaria during pregnancy in all settings.

Managing anaemia in pregnancy – Anaemia is a common, and potentially dangerous, complication of malaria in pregnancy. Anaemia prevention and management should be a priority when working with pregnant women in malaria-endemic areas during humanitarian emergencies.

- For prevention, 1 tablet containing ferrous sulfate, 200 mg (equivalent to 60 mg elemental iron), plus folic acid, 400 µg, should be given daily throughout pregnancy;
- In areas of moderate-to-high malaria transmission in sub-Saharan Africa, where there is a high risk of asymptomatic malaria infection in pregnancy, pregnant women should receive intermittent preventive treatment with sulfadoxine-pyrimethamine (see Intermittent preventive treatment in pregnancy [*IPTp*] below);
- Presumptive treatment for intestinal worms (using a single 500 mg dose of mebendazole or 400 mg dose of albendazole) can be given once during the second or third trimester of pregnancy, but these drugs must not be given in the first trimester.

Intermittent preventive treatment in pregnancy (IPTp) – In areas of moderate-to-high falciparum transmission in sub-Saharan Africa, IPTp with sulfadoxine-pyrimethamine (provided to pregnant women through ANC services) can reduce the incidence of placental infection, anaemia, and low birth-weight. It is not feasible to implement this during the acute phase of an emergency unless ANC services already exist.

Currently, the only drug for which there is documented evidence of safety and effectiveness when used as IPTp is sulfadoxine-pyrimethamine (SP). SP given at each scheduled antenatal care visit after the first trimester or quickening (first fetal movement perceived by the mother) has proven effective in reducing placental infection and material parasitaemia, preventing maternal anaemia, and improving birth-weight.

Key points on IPTp of semi-immune pregnant women in areas of high malaria endemicity are listed in Table 6.8.

Non-falciparum malaria in pregnancy is treated the same way as for non-pregnant patients, with the exception of radical cure of *P. vivax* and *P. ovale*, for which primaquine is contraindicated in pregnancy (see *Managing non-falciparum malaria*).

Uncomplicated falciparum malaria in pregnancy – All pregnant women with confirmed malaria should receive urgent treatment with effective antimalarials. Symptoms in partially-immune pregnant women can be mild. ACTs should be used in the 2nd and 3rd trimesters. Quinine plus clindamycin can be used safely for treatment in any trimester. An ACT is indicated in the first trimester if this is the only treatment immediately available, if treatment with 7-day quinine plus clindamycin fails, or if there is uncertainty of compliance with a 7-day treatment. Primaquine, tetracyclines and doxycycline are contraindicated in pregnancy (see Annex VI).

Table 6.8 **IPTp in areas of high malaria endemicity**

Key points
• Give IPT with SP (3 tablets) for all pregnant women **at each scheduled antenatal care visit after the first trimester**. WHO recommends a schedule of four antenatal care visits. — The first IPTp-SP dose should be administered as early as possible during the 2nd trimester[a] of gestation. — The last dose of IPTp with SP can be administered up to the time of delivery, without safety concerns. — IPTp should ideally be administered as directly observed therapy (DOT). — SP can be given on an empty stomach or with food. • IPTp should be given under direct observation at the time of the ANC visit.
Precautions
• Do *not* give IPT with SP more frequently than once a month. If a woman develops malaria after receiving SP for IPTp she should be treated with an antimalarial that does not contain SP. • Do *not* give SP to women who have a history of allergy to sulfa drugs. If an allergic reaction is suspected after the first IPT dose, do not give further doses. • Do *not* give IPT with SP to PLHIV who already receive cotrimoxazole prophylaxis. • Advise women to return for the next dose of IPT after a clearly defined time interval (e.g. 4 weeks). • Explain to women that they can still get clinical malaria, despite IPT, and that they should return immediately to the clinic if they develop a fever or anaemia.

[a] IPTp administration should be avoided during the 1st trimester of gestation but should start as soon as possible in the 2nd trimester. The fact that a woman has entered the second trimester can be determined by the onset of quickening or by measurement of fundal height by ANC health personnel.

Severe malaria in pregnancy – Pregnant women with severe malaria should receive urgent medical care, because of the high risk of maternal and fetal mortality. Parenteral artesunate is the treatment of choice for severe malaria; if this is not available artemether is preferable to quinine in late pregnancy as quinine is associated with 50% risk of hypoglycaemia. Parenteral treatment should be given for at least 24 hours, or until the woman can tolerate oral medication, and should always be followed by a full course of ACTs. See Annex VI for indications and dosages.

Hypoglycaemia, acute pulmonary oedema, hyperpyrexia, postpartum haemorrhage, premature delivery and perinatal death are particular risks. Presentation and treatment of these complications include:

- *Hypoglycaemia* is a significant risk for all pregnant women with malaria. The risk is greatest in the 2nd and 3rd trimesters, may be present on admission, and can be a complication of quinine treatment. The increased risk of hypoglycaemia persists into the post-partum period and blood glucose should be monitored every 4 hours. In women with cerebral malaria, hypoglycaemia may be asymptomatic or may cause deterioration in the level of consciousness, extensor posturing, or convulsions. Differential diagnoses include sepsis, meningitis and eclampsia. Hypoglycaemia may recur after correction with IV glucose, and blood glucose should be monitored frequently.
- *Convulsions* in pregnant women can be caused by hypoglycaemia, cerebral malaria, or eclampsia. An appropriate evaluation should be done to determine the cause.
- *Acute pulmonary oedema* is particularly common during labour and immediately after delivery. Severe anaemia and the increase in blood volume and peripheral resistance that follows placental separation may precipitate acute pulmonary oedema and heart failure. This is a medical emergency that requires immediate treatment:
  - Check for increased respiratory rate, chest signs (crackles on auscultation), and hepatomegaly;
  - If pulmonary oedema is suspected, position the patient upright, give high concentration oxygen, and IV furosemide, 0.6 v/kg (adult dose 40 mg);
  - If pulmonary oedema is associated with overhydration, stop all intravenous fluids and give furosemide (same dose as above).
- *Severe anaemia.* Blood transfusion is indicated in women with erythrocyte volume fraction (EVF = haematocrit) lower than 20% or a haemoglobin concentration less than 7 g/dl:

— Give screened packed red blood cells by slow transfusion over 6 hours and furosemide (frusemide) 20 mg intravenously;
— Folic acid and iron supplements may be required during recovery.

Malnourished persons

Malnutrition is a significant cause of morbidity and mortality in humanitarian emergencies, and often coexists with malaria. *There are different ways of classifying malnutrition*. The most common include differentiating between overnutrition and undernutrition, distinguishing chronic from acute forms, and within acute forms, distinguishing moderate from severe forms. Overnutrition remains a lesser issue in humanitarian contexts. Undernutrition can result in acute malnutrition (i.e. wasting or *low weight-for-height*, nutritional oedema), chronic malnutrition (i.e. stunting or *low height-for-age*), micronutrient malnutrition, and inter-uterine growth restriction.

The focus in humanitarian emergencies is usually acute malnutrition and micronutrient deficiencies, as they manifest most rapidly and visibly. Underweight (i.e. *low weight-for-age*), a composite measure of acute and chronic malnutrition, is an important marker in emergency contexts. Chronic malnutrition and underweight reflect underlying nutritional vulnerability, and emergency-prone populations are more likely to be chronically malnourished. Acute malnutrition is further divided into severe (SAM) and moderate (MAM) forms, with global acute malnutrition (GAM) encompassing both.

Diagnosis – Patients with SAM may have an asymptomatic malaria infection and should therefore be screened for malaria (using microscopy or RDT) on admission to a therapeutic feeding programme and weekly thereafter until discharge.

Treatment – Once a patient has tested positive for malaria, an ACT should be given for treatment. There are limited data on the effect of malnutrition on chloroquine, doxycycline, quinine, sulfadoxine-pyrimethamine and tetracycline, and insufficient evidence to suggest that dosages (in mg/kg body-weight) of any antimalarial should be changed in patients with malnutrition.

Severe malaria – Patients with severe malaria and SAM are at very high risk of death and require intensive medical and nursing care. They should be hospitalized in a therapeutic feeding centre and treated with an effective parenteral artemisinin derivative.

People living with HIV

There is considerable geographic overlap between malaria and HIV, indicating the potential for substantial numbers of co-infected individuals in many humanitarian emergencies. Worsening HIV-related immunosuppression may lead to more severe manifestations of malaria. In HIV-infected pregnant women, adverse effects of placental malaria on birth-weight are increased. HIV-infected individuals may suffer more frequent and higher density infections or an increased risk of severe malaria and malaria-related death, depending on transmission intensity and level of malaria immunity.

Treatment – Information is still limited on how HIV infection modifies therapeutic responses to ACTs or on interactions between antimalarial drugs and antiretrovirals. Thus, PLHIV who develop malaria should receive the appropriate antimalarial regimen for their setting, though sulfadoxine-pyrimethamine should be avoided by those taking cotrimoxazole prophylaxis and amodiaquine should be avoided by PLHIV taking zidovudine or efavirenz. Given the potential risks of treatment interactions, particular focus for PLHIV should be on malaria prevention (e.g. usage of LLINs) where feasible.

Displaced and returnee populations

Generally, mass population movements in geographical areas of mixed endemicity can increase the risk of severe malaria epidemics, especially when people living in an area with little or no malaria transmission move to an endemic area (e.g. displacement from highland to lowland areas). The lack of protective immunity, the concentration of people in exposed settings, the breakdown in public health and preventive activities, together with difficulties in accessing effective treatment and with concomitant infections and malnutrition, all render populations vulnerable to epidemic malaria. Such circumstances are also ideal for the development of parasite resistance to antimalarials. For these reasons, particular efforts must be made to deliver, free-of-charge, prompt diagnostic testing and effective antimalarial treatment to populations at risk.

References

- WHO/FAO (2004). *Vitamin and mineral requirements in human nutrition.* 2nd ed. Geneva, World Health Organization.
- WHO (2005c). *Handbook: IMCI integrated management of childhood illness.* Geneva, World Health Organization.

- WHO (2009). *Acute care integrated management of adolescent and adult illness (IMAI) Guidelines for first-level facility health workers at health centre and district outpatient clinic*. Geneva, World Health Organization.
- WHO (2010). *Guidelines for the treatment of malaria 2nd ed.* Geneva, World Health Organization. http://www.who.int/malaria/publications/atoz/9789241547925/en/index.html
- WHO (2011). *Good practices for selecting and procuring rapid diagnostic tests for malaria*. Geneva, World Health Organization. http://www.who.int/malaria/publications/atoz/9789241501125/en/index.html
- WHO (2011b). *Universal access to malaria diagnostic testing: an operational manual*. Geneva, World Health Organization. http://www.who.int/malaria/publications/atoz/9789241502092/en/index.html
- WHO (2012). *Malaria Rapid Diagnostic Test Performance. Results of WHO product testing of malaria RDTs: 4 (2012)*. Geneva, World Health Organization. http://www.who.int/malaria/publications/rapid_diagnostic/en/index.html

Finding out more

- WHO/UNICEF/UNU (2001). *Iron deficiency anaemia assessment, prevention, and control: a guide for programme managers*. Geneva, World Health Organization. http://www.who.int/nutrition/publications/micronutrients/anaemia_iron_deficiency/WHO_NHD_01.3/en/index.html
- WHO (2005). *Guiding principles for feeding non-breastfed children 6–24 months of age*. Geneva, World Health Organization. http://www.who.int/maternal_child_adolescent/documents/9241593431/en/index.html
- WHO (2005b). *Guiding principles for complementary feeding of the breastfed child*. Geneva, World Health Organization. http://www.who.int/maternal_child_adolescent/documents/a85622/en/index.html
- WHO (2011c). *Haemoglobin concentrations for the diagnosis of anaemia and assessment of severity*. Vitamin and Mineral Nutrition Information System. Geneva, World Health Organization.
- WHO (2011d). *Guidelines: Intermittent iron supplementation in preschool and school-age children*. Geneva, World Health Organization. http://www.who.int/nutrition/publications/micronutrients/guidelines/guideline_iron_supplementation_children/en/index.html
- WHO (2011e). *Guidelines: Intermittent iron and folic acid supplementation in menstruating women*. Geneva, World Health Organization. http://www.who.int/nutrition/publications/micronutrients/guidelines/guideline_iron_folicacid_suppl_women/en/index.html

- WHO (2012c). *Guidelines: Intermittent iron and folic acid supplementation in non-anaemic pregnant women*. Geneva, World Health Organization. http://www.who.int/nutrition/publications/micronutrients/guidelines/guideline_intermittent_ifa_non_anaemic_pregnancy/en/index.html
- WHO (2012d). *Guidelines: Daily iron and folic acid supplementation in pregnant women*. Geneva, World Health Organization. http://www.who.int/nutrition/publications/micronutrients/guidelines/daily_ifa_supp_pregnant_women/en/index.html
- WHO (2013). *Management of severe malaria: a practical handbook – 3rd ed*. Geneva, World Health Organization. http://www.who.int/malaria/publications/atoz/9789241548526/en/index.html

CHAPTER 7
Prevention

This chapter
- discusses available methods for malaria prevention in humanitarian emergencies
- outlines how to choose prevention interventions and organize activities
- describes operational aspects of mosquito biology and behaviour

Malaria prevention in humanitarian emergencies

While the first priorities in the acute phase of an emergency are prompt and effective diagnosis and treatment of people with malaria, prevention can make an important contribution to reducing the risk of infection and saving lives. This chapter addresses prevention related to humanitarian emergencies, including vector control, personal protection against mosquito bites, intermittent preventive treatment (IPT), and the prospects for a malaria vaccine. Chapter 5 addresses prevention specific to malaria outbreaks.

The same prevention methods are usually appropriate for both human-induced and natural disasters and the same constraints often apply. In emergencies with a risk of malaria, the most important initial questions related to prevention are:

- Is malaria prevention likely to be useful (e.g. how great is the malaria burden? Where is transmission occurring? What is the risk and does it necessitate pro-active rather than reactive response?)?
- Which malaria prevention methods are most feasible (e.g. depending on local vector species, access, security, population mobility, human resources, funding, and logistics)?

A situation assessment (see Chapter 3), particularly regarding malaria control strategies in both the host country and the country of origin for refugee populations, should answer some of these questions. However, the nature of humanitarian emergencies presents both opportunities and constraints that may justify different prevention approaches (see Table 7.1).

Table 7.1 **Factors that may affect prevention approaches in humanitarian emergencies**

Opportunities
• Readily available trained workforce and adequate supervision. • Concentrated population, potentially facilitating access, health education and logistics. • Additional humanitarian funding provided by international donors.
Constraints
• Access to affected populations may be limited by conflict or geography. • Priority for curative care and food security may limit funding and human resources available for vector control. • Baseline information on vector presence and insecticide susceptibility may be missing. • Most malaria transmission may occur outside resettlement sites. • Language barriers may make it difficult to provide information. • Temporary shelters may differ from local housing and make IRS and use of LLINs more problematic. • Unstable and unpredictable situations make longer-term planning difficult. • Evaluation of prevention strategies can be difficult. • Problems in securing orders and importing commodities (e.g. insecticide, pumps, LLINs) quickly enough.

The evidence base for effective vector control in humanitarian emergencies has expanded and field experience confirms the feasibility of timely vector control interventions. New tools for vector control and personal protection have the potential to significantly improve delivery of timely interventions and make a difference in humanitarian emergencies in a variety of contexts (see Chapter 9). When planning vector control, it is important to consider how malaria prevention approaches will need to change as the humanitarian emergency moves from acute to post-acute phase or consolidates into a chronic emergency.

- **Acute-phase** priorities are prompt and effective diagnosis and treatment of all clinical malaria episodes to limit avoidable malaria deaths. Where feasible, this should be supplemented with barrier methods of mosquito-bite prevention – most commonly LLINs – aimed first at priority vulnerable groups with the highest risk of developing severe malaria but aiming to cover all at risk (i.e. universal coverage). Alternatively, IRS may be used in some contexts. IRS requires substantial planning, logistics and human and material resources, but has been shown to protect people in well-organized settings such as transit camps. IRS is not suitable for protecting individual households scattered over large distances or where shelters are very temporary (i.e. less than about three months) and without surfaces that can be readily sprayed (e.g. very open structures of sticks with a tarpaulin roof). IRS is best suited for protecting larger populations in more compact settings, where shelters are more permanent and solid.

- **Post-acute phase** priorities broaden as the situation stabilizes. It may be possible to prevent new infections by adding high-coverage community vector control measures i.e. covering a high percentage of the population with insecticide-treated nets (LLINs) or spraying a high percentage of dwellings with residual insecticide to achieve an impact on transmission.

During the acute phase, decisions on vector control and prevention will depend on:

- — malaria infection risk;
- — behaviour of the human population (e.g. mobility, where they are sleeping or being exposed to vector mosquitoes);
- — behaviour of the local vector population (e.g. indoor resting, indoor biting, early evening or night biting);
- — the type of shelter available (e.g. ad-hoc refuse materials, plastic sheeting, tents, more permanent housing).

New vector control and personal protection tools have been specifically designed for acute-phase emergencies. For example, in new settlements where shelter is very basic, insecticide-treated plastic sheeting may be more appropriate, acceptable and feasible than LLINs or IRS. Insecticide-pyrethroid-treated plastic sheeting should not be used in areas where the local malaria vectors are resistant to pyrethroids. There are not currently sufficient data on the efficacy and safety of insecticide-treated plastic sheeting for a formal WHO recommendation. However, operational realities may necessitate such approaches when neither IRS nor LLINs are operationally feasible. It is for these reasons that they are included in this handbook. Other methods, including permethrin-treated blankets and topical and spatial repellents, may sometimes be useful as adjunct control measures.

As the situation stabilizes, longer-term approaches can be introduced. Prevention methods may change as the population becomes less mobile and temporary shelters are replaced by more permanent structures. There may be greater opportunities for providing LLINs. LLINs are widely used for malaria prevention and most communities in endemic areas will have some exposure to them. IPT (see Chapter 6) may be useful during post-acute or chronic emergencies in appropriate settings in sub-Saharan Africa as it requires relatively few resources compared to LLINs or IRS.

Long-lasting insecticidal nets (LLINs)

Whenever possible, LLINs should be provided in sufficient numbers to cover everyone exposed to transmission in target communities. Normally, a combination of campaign (catch-up) and routine (keep-up) distribution systems are needed to sustain coverage. When supplies are constrained, LLINs can still provide personal protection to risk groups (e.g. young children and pregnant women) in high transmission areas. LLINs may be distributed through specialized delivery channels (e.g. emergency mass distributions) or integrated strategies (e.g. through antenatal care, vaccination). Only WHOPES-approved LLINs should be used. While the insecticide on an LLIN should last for at least three years, the physical lifespan of an LLIN is highly variable, thus requiring continuous distribution and monitoring in harsh emergency settings. Recent data from control programmes (not operating under emergency circumstances), suggest that the life span of many LLINs may be less than 24 months.

LLINs are highly effective in the control of malaria in most settings and are accepted by malaria-affected communities worldwide. Sleeping under an untreated mosquito net provides a physical barrier against mosquitoes; however, mosquitoes can still bite through any tears or holes or if any part of the body is touching an untreated net. Nets treated with a pyrethroid insecticide provide a significantly increased level of protection. Untreated nets that require insecticide treatment are not practical in emergencies, and it is preferable to deploy WHO Pesticide Evaluation Scheme (WHOPES)-approved LLINs. These are factory-treated mosquito nets that retain their biological activity for a minimum period of time and number of washes, currently at least 20 standardized WHO washes and 2–3 years of recommended use under normal field conditions.

Pyrethroids are the only class of chemical currently approved for use on mosquito nets. They are very safe for humans, but toxic in low doses for mosquitoes – killing or repelling mosquitoes before they can enter the net or bite those sleeping under it. Pyrethroid treatment of nets provides personal protection for those who sleep under them even when coverage is low. High coverage of LLINs may also provide a mass impact (i.e. provide some protection for those not sleeping under LLINs) through reduction of mosquito populations and reduced longevity of adult mosquitoes. Insecticide-treated nets have a killing or disabling effect on other biting vectors (e.g. lice, fleas, ticks, sandflies). Bedbugs, however, often exhibit a higher degree of tolerance to insecticides and may not be killed (e.g. some IRS programmes noted that bedbugs are simply agitated by insecticides and appear during the day, upsetting house owners).

WHOPES-approved LLINs are made of polyester, polyethylene or polypropylene mesh and meet the criteria for "long-lasting" labelling of containing insecticide that is wash-resistant and can remain efficacious even when the net has been washed 20 times. The physical durability of netting material is an important quality that varies with the *denier*, or weight of the fibre, and other net characteristics. Nets with a denier value less than 100 are not recommended due to relative ease of tearing. The contexts in which LLINs are used may also determine their physical durability (e.g. if they are subject to physical abrasion, tearing by reed sleeping mats, chewing by rodents, or fire damage).

WHO has produced protocols for assessing the longevity of LLINs under field conditions that are available on WHO Global Malaria Programme and WHOPES websites. Table 7.2 provides a list of WHO recommended LLINs as of October 2012. The WHOPES website should be consulted for the latest update at http://www.who.int/whopes/en/.

General preconditions for successful LLIN distribution during emergencies – People must be under their LLINs at the time of night that most vector mosquitoes bite. Most people are willing to use LLINs, and behaviour-change communications (see Chapter 8) can improve population usage. It is important to organize and implement an effective community promotion campaign prior to distribution.

LLINs can be used indoors or outdoors, wherever the population normally sleeps, but appropriate installation and support mechanisms should be supplied and explained (Figure 7.1). Rigid or durable dwellings (e.g. mud huts) are suitable for hanging LLINs. Tent structures present a challenge, as some may not be of sufficient size or construction to be suitable for hanging LLINs; some tents can be sprayed with a residual insecticide instead, but duration of effect depends on the compound used and the type of tent material. Consultation with an experienced field entomologist or vector control expert is advised in such circumstances.

Good logistics are crucial. Ensure the delivery system is adequate and that information on population size and location of houses and population can be acquired. Timely procurement, factoring in shipping needs, storage capacity, and local transport will enable LLINs to be made available when and where needed in sufficient numbers and without gaps in the supply chain. Where possible, LLINs should be ordered well in advance of planned distribution, to ensure timely delivery and standard transportation costs. LLINs of standard specifications (e.g. double, family-size) are faster to procure, as manufacturers tend to maintain stocks. In chronic emergency situ-

Table 7.2 **WHO recommended LLINs**

Product name	Product type	Status of WHO recommendation	Status of publication of WHO specification
DawaPlus® 2.0	Deltamethrin coated on polyester	Interim	Published
Duranet®	Alpha-cypermethrin incorporated into polyethylene	Interim	Published
Interceptor®	Alpha-cypermethrin coated on polyester	Full	Published
Lifenet®	Deltamethrin incorporated into polypropylene	Interim	Published
MAGnet™	Alpha-cypermethrin incorporated into polyethylene	Interim	Published
Netroprotect®	Deltamethrin incorporated into polyethylene	Interim	Published
Olyset®	Permethrin incorporated into polyethylene	Full	Published
Olyset® Plus	Permethrin and PBO incorporated into polyethylene	Interim	Pending
Permanet® 2.0	Deltamethrin coated on polyester	Full	Published
Permanet® 2.5	Deltamethrin coated on polyester with strengthened border	Interim	Published
Permanet® 3.0	Combination of deltamethrin coated on polyester with strengthened border (side panels) and deltamethrin and PBO incorporated into polyethylene (roof)	Interim	Published
Royal Sentry®	Alpha-cypermethrin incorporated into polyethylene	Interim	Published
Yorkool® LN	Deltamethrin coated on polyester	Full	Published

Notes:
1. Reports of the WHOPES Working Group Meetings should be consulted for detailed guidance on use and recommendations. These reports are available at http://www.who.int/whopes/recommendations/wgm/en/); and
2. WHO recommendations on the use of pesticides in public health are valid ONLY if linked to WHO specifications for their quality control. WHO specifications for public health pesticides are available at http://www.who.int/whopes/quality/newspecif/en/.

ations with less time pressure, it is important to determine the most suitable LLINs with regard to shape, size, and colour. LLINs are best distributed before the start of the malaria season to provide maximum transmission reduction. However, because LLINs will last for years, distribution will be effective whenever it can be implemented.

Planning and procurement – It is important to verify existing LLIN coverage (e.g. that there has been no recent distribution of new undamaged LLINs). A pre-distribution survey is ideal, though not always feasible in an emergency context. In cases where good coverage already exists, additional tools might still be considered for distribution (e.g. impregnated blankets).

Beneficiaries must be clearly defined and the highest possible coverage aimed for in the shortest time. For example, 'catch-up' mass distribution can aim for full household coverage of 100% of target populations. Alternatively, if only limited numbers of LLINs are available, initial distribution should target high-priority vulnerable groups through 'keep-up' targeted distribution (e.g. pregnant women and children under 5 through antenatal clinics, routine immunization programmes, feeding centres or vaccination campaigns).

In planning procurement quantities, aim to distribute enough LLINs to achieve 100% coverage, with one LLIN for two persons. A good way to achieve this at household level is to distribute LLINs per household at a rate of one LLIN for every two household members, rounding up in households with odd numbers of members. The procurement ratio must be adjusted to allow for this rounding up. For example, this implies a procurement ratio of 550 LLINs per 1000 population (i.e. one LLIN for 1.8 people) in a population with a mean household size of five.

Colour and shape are important considerations. LLINs are available in rectangular and conical shapes and several colours. Conical nets are more expensive, but fit better in conical dwellings and are preferred in some settings (e.g. Somalia). Some colours may be refused or misused by beneficiaries because of the cultural significance of certain colours (e.g. a connection might exist between white colour and death). White LLINs show dirt and may get washed more often than coloured nets. For outside sleepers, impregnated hammock nets might be more appropriate (e.g. Cambodia). Opaque cotton sheeting (e.g. Dumuria) is preferred in some settings, such as South Sudan, because it provides more privacy but does not retain insecticides as long as other net fibres. In emergencies, the specifications available in stock are preferable in order to facilitate rapid deployment.

Distribution – Define how and where LLINs will be distributed (see Table 7.3). A successful approach used in emergencies is mass distribution to beneficiaries who are invited by the community leaders to a central location and given nets. Another successful approach is "hang-up and use" distribution, in which LLINs are installed on site by distributors. This labour intensive approach is slower, but may have a positive impact on the proper retention and use of each LLIN. Handling large numbers of new LLINs can result in temporary rashes or other allergic reactions, due to the build-up of pyrethroid on the net's surface during storage, and as such, staff should use protective clothing (e.g. gloves, long sleeves).

To reduce the risk of beneficiaries re-selling LLINs, each LLIN should be removed from its packaging, marked with a number or name, and reception signed for by the head of household. Some LLINs are already printed with an individual number, which helps with monitoring and evaluation of distribution.

Table 7.3 **LLIN distribution by phase of emergency and level of transmission**

Acute-phase emergencies in high/moderate transmission areas
Ideally, universal LLIN coverage of the whole community should be the objective. If insufficient LLINs are stockpiled, prioritize highest risk groups: 1. all beds/patients in hospitals and therapeutic feeding centres (TFC), and households of TFC patients on discharge; 2. pregnant women and children under 5 years of age; 3. populations living in areas of high transmission (so-called "hot spot" transmission zones).
Acute-phase emergencies in low-transmission areas
Use LLINs only in clinical settings (e.g. TFC beds, hospital beds), with no community distribution schemes.
Post-acute or chronic emergencies in high/moderate transmission areas
Universal coverage is ideal. If resources do not allow this: 1. target distribution to areas and populations most at risk; 2. provide full LLIN coverage to households with pregnant women or children under 5 years of age, with catch-up distribution schemes through antenatal care (ANC), immunization programmes and primary health care; 3. provide LLINs to People Living with HIV (PLHIV), who are more at risk of developing severe malaria.
Post-acute or chronic emergencies in low-transmission areas
Select intervention according to local practice. LLINs are not the only choice.

Health communications – LLIN distribution campaigns should be accompanied by comprehensive behaviour change communication approaches (e.g. drama, radio) to encourage appropriate usage and maintenance and improve intervention success. While increasing evidence from most settings in Africa indicates net use is high, with up to 80% of nets being used (WHO, 2010), they are sometimes diverted for other uses such as fishing, clothing (e.g. wedding dresses), or crop protection. These alternative uses should be discouraged and safety should be emphasized in promotion campaigns (e.g. no fire or lit cigarettes near LLINs) as should appropriate usage (e.g. ensuring LLINs are tucked in and are used every night even where mosquitoes are not heard or seen). LLINs used by anyone with a highly contagious disease (e.g. viral haemorrhagic fever) must be disinfected, disposed of appropriately and replaced. Each LLIN must be accompanied by simple country-specific information (e.g. leaflets) to enable people to use them correctly. These should include pictorial information on how to hang the nets (e.g. ropes and nails when suspended indoors from walls or ceilings, rounded poles for outdoor LLIN hanging – to avoid damaging LLINs, and UV radiation considerations if LLINs stay outdoors during the day). Figure 7.1 provides an example from Afghanistan.

Monitoring – Coverage should be assessed by monitoring at distribution, and then at one and six months, to determine appropriate use and retention. Standard indicators from a management information system (MIS) or Demographic and Health Survey (DHS) should be used. If standardized indicators are not available, appropriate indicators for inclusion are:

- Coverage = number of LLINs distributed/target population size (%);
- Usage rate = number of people using LLINs/number of people given LLINs (%);
- Retention rate = number of people retaining LLINs/number of people originally given LLINs (%);Net failure rate = average number of holes per LLIN following WHO criteria as given in the WHO guidelines on monitoring LLIN durability. Indicators of survivor, attrition, and failure of LLINs are defined further in http://whqlibdoc.who.int/publications/2011/9789241501705_eng.pdf;
- Behaviour change indicator (e.g. proportion of population that can recall or understand message).

Figure 7.1 **Example approaches to hanging LLINs indoors and outdoors**

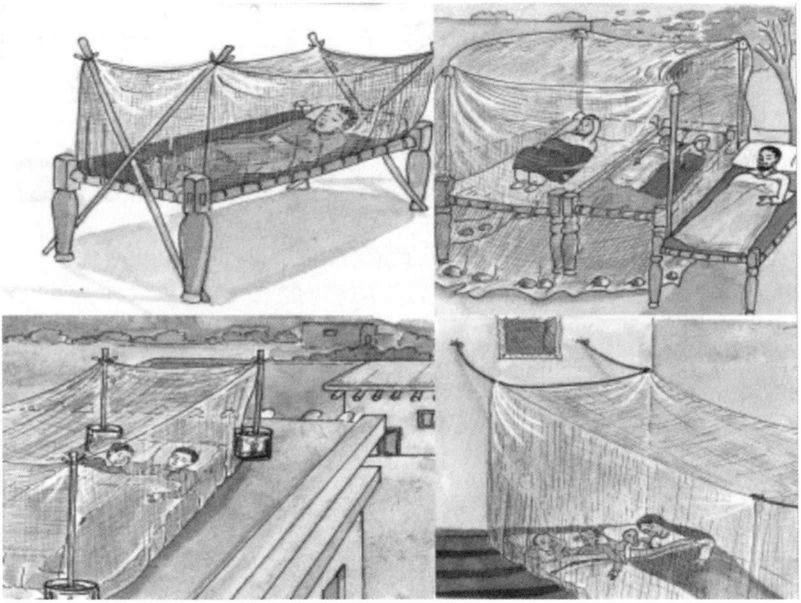

Source: HealthNet-TPO

Indoor residual spraying (IRS)

IRS, when implemented properly, is a highly-effective intervention that provides protection to entire communities through rapidly impacting vector populations – reducing densities and longevity of vectors and their capacity to transmit malaria parasites. IRS effectiveness is dependent on the quality of spraying operations (i.e. at least 80% of premises in target communities must be properly sprayed). IRS is usually effective for 3–6 months, depending on the insecticide used, type of surface sprayed, and transmission seasonality; more information may be obtained at: http://www.who.int/whopes/en/insecticides_IRS_malaria_09.pdf.

IRS involves spraying a residual insecticide onto the inside walls and ceilings or the underside of the roof and eaves of houses, in order to kill mosquitoes when they come indoors to feed and rest. The basic preconditions for IRS are:

- an endophilic malaria vector (i.e. mosquitoes that enter and rest inside houses long enough to come in contact with the insecticide);

- an insecticide to which the vector is susceptible;
- a population living in dwellings with sprayable surfaces (e.g. huts made of branches with wide voids or gaps will be difficult to spray);
- a population that is not nomadic and structures that are permanent for the coming 6–12 months;
- willingness of the population to accept IRS;
- communities living in close proximity (e.g. villages, towns, transitional camps). IRS is not the correct tool for dispersed rural populations, as logistics become increasingly difficult;
- capacity and training to deliver a good-quality application of insecticides, as low-quality IRS campaigns may cause more harm than benefit.

To be effective as a community control measure, IRS requires coverage of at least 80% of dwellings, ensuring that the majority of mosquitoes are exposed to the insecticide. Indoor sleeping is not strictly a requirement for IRS success. For example, in Afghan refugee settlements people sleep outdoors on hot summer nights, but indoor spraying still protects against malaria because mosquitoes rest indoors during the day.

Given the operational requirements, IRS is most suited to areas that are somewhat stable politically, where advance planning and close monitoring of campaign coverage are possible. In acute-phase emergencies, there may be opportunities for IRS during planning for refugee transit camps and other well-organized settings. An IRS campaign requires substantial advance planning, good logistics support, and human and material resources. Its effectiveness is highly dependent on operational factors – timely delivery of commodities, on-site expertise and capacity, good organization and planning, trained staff, supervision, and appropriate health communications.

Timing is critical in areas of seasonal transmission and at least 80% of shelters (e.g. houses, temporary shelters, hospitals, orphanages, schools) must be sprayed with insecticide before onset of the expected peak transmission season.

Good-quality compression sprayers that comply with WHO specifications must be used for IRS. Compression sprayers are prone to wear and tear and may need to be imported, together with spare parts and insecticides. Importation requires time and adequate planning. Proper maintenance is important to ensure that compression sprayers are efficient and their working life is maximized. IRS campaigns have failed due to neglect of spray equipment (e.g. pumps, nozzles and lances becoming blocked through lack of daily cleaning or maintenance). Information on WHO specifications for spray equipment can be found at the WHOPES website: http://www.who.int/whopes/equipment/en/.

Figure 7.2 **Example compression sprayer**

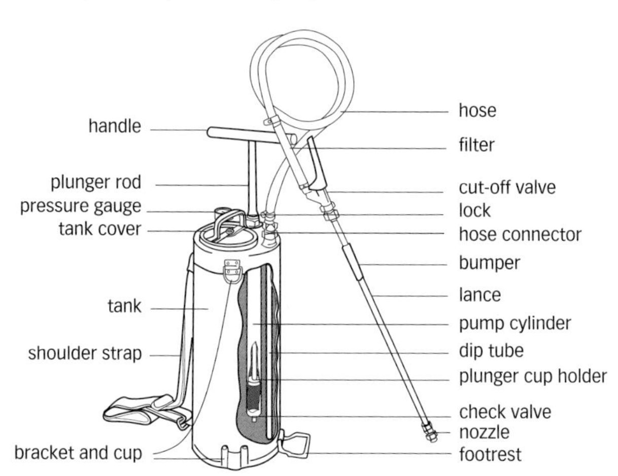

Source: WHOPES

To achieve the desired impact, IRS campaigns must be implemented properly. Insecticide must be applied safely, and operations properly supervised. Spray teams require at least two days of intensive theoretical and practical training before they can start field operations and these operations must be closely supervised and monitored. WHO produced an illustrated manual that can be used for training spray teams and maintaining spraying equipment (http://www.who.int/whopes/equipment/en/). In addition, WHO/GMP has recently developed an operational manual for IRS for malaria transmission control and elimination, available on the WHO website at http://www.who.int/malaria/publications/atoz/9789241505123/en/index.html. This manual describes policy and strategic issues – including how to manage an IRS programme and conduct a house spray.

Important instructions to be followed by sprayers when implementing indoor residual spraying are summarized in Table 7.4.

Table 7.5 lists the insecticides commonly used for IRS, recommended application rates (g/m^2), and residual life or persistence. Any WHO-recommended insecticides can be used if available locally and known to be effective. WHO specifications for public health pesticides – for quality control and international trade – are available at http://www.who.int/whopes/quality. IRS is governed by country registration regulations. There have been a few exceptions in emergencies, where WHOPES-approved

Table 7.4 **Household aspects of indoor residual spraying**

Household preparations
• Inform household occupants of the spraying schedule and reasons for spraying. • Allow occupants time to prepare and vacate living areas. • Remove all household items, including water, food, cooking utensils and toys. Items that cannot be removed should be carefully covered. • Move pets and domestic animals away from the home. • Occupants *must* leave before spraying begins. • Areas occupied by sick people who cannot be moved must not be sprayed.
Household procedures after spraying
• Advise occupants to stay outside for at least one hour while sprayed surfaces are drying. • Instruct occupants to sweep or mop the floor before children and pets are allowed to re-enter. • Instruct occupants not to wipe or clean sprayed surfaces.
Waste disposal
• At the end of each day's spraying, pumps should be cleaned by the standard 'triple rinse' method so that waste water is contained. • Never pour insecticide into rivers, pools or drinking-water sources. • Never reuse or burn empty insecticide containers.

products have been considered without full national registration (e.g. Liberia, Chad). In areas with high LLIN usage, pyrethroids should not be used for IRS to avoid rapid selection of pyrethroid resistance. The use of non-pyrethroid insecticides should be based on the susceptibility status of the local vectors to those insecticides. New long-lasting formulations of non-pyrethroid alternatives are being evaluated by WHO and may in the future ease the overreliance on pyrethroids for IRS. Refer to the WHO Global Plan for Insecticide Resistance Management for more information http://www.who.int/malaria/publications/atoz/gpirm/en/index.html.

If malaria transmission is seasonal, it is important to spray just before or at the start of the season. In an outbreak, IRS will only achieve its full impact if walls are sprayed well before the epidemic peak. This requires detailed advance planning and considerable logistic capacity and is rarely possible in practice in emergency situations.

Table 7.5 **WHO-recommended insecticides for IRS against malaria vectors**

Insecticide compounds and formulations[a]	Class group[b]	Dosage (g a.i./m²)	Mode of action	Duration of effective action (months)
DDT WP	OC	1–2	contact	>6
Malathion WP	OP	2	contact	2–3
Fenitrothion WP	OP	2	contact and airborne	3–6
Pirimiphos-methyl WP & EC	OP	1–2	contact and airborne	2–3
Bendiocarb WP	C	0.1–0.4	contact and airborne	2–6
Propoxur WP	C	1–2	contact and airborne	3–6
Alpha-cypermethrin WP & SC	PY	0.02–0.03	contact	4–6
Bifenthrin WP	PY	0.025–0.05	contact	3–6
Cyfluthrin WP	PY	0.02–0.05	contact	3–6
Deltamethrin WP, WG	PY	0.02–0.025	contact	3–6
Etofenprox WP	PY	0.1–0.3	contact	3–6
Lambda-cyhalothrin WP, CS	PY	0.02–0.03	contact	3–6

[a] CS: capsule suspension; EC= emulsifiable concentrate; SC= suspension concentrate; WG= water dispersible granule; WP= wettable powder.

[b] OC Organochlorines; OP= Organophosphates; C= Carbamates; PY= Pyrethroids.

Note: WHO recommendations on the use of pesticides in public health are valid ONLY if linked to WHO specifications for their quality control. WHO specifications for public health pesticides are available on the WHP home page at http:/www.who.int/whopes/quality/en/.

Source: WHOPES (http:/www.who.int/whopes/Insectisides_IRS_Malaria_09.pdf

IRS programme monitoring

Appropriate indicators for inclusion in quality and effectiveness monitoring are:

- Coverage = number of structures sprayed / number of structures in the target areas (%);
- Amount of insecticide used per structure = amount of insecticide used / number of structures sprayed (%). This is one measure of the efficiency and correct usage of the insecticide;
- Maintenance = % of pumps that were correctly maintained and functioning well at the end of the campaign;
- Acceptability = % of households that refused to have their homes sprayed;
- Acceptability after IRS = % heads of households who complained after a spray campaign.

Chemoprevention

Chemoprevention, through periodic intake of antimalarials approved for this purpose, can provide protection against malaria and its adverse consequences during pregnancy, infancy and early childhood (see Chapter 6). Chemoprevention may be particularly useful during the post-acute phase or during chronic emergencies in sub-Saharan Africa, as it requires minimal specialist knowledge and can be administered by health staff.

Intermittent preventive treatment (IPT) is one form of chemoprevention that involves repeated delivery of a full treatment dose of an antimalarial drug – usually sulfadoxine-pyrimethamine (SP) – to a particular risk group to prevent the consequences of malaria infection. Treatments are given to asymptomatic people at pre-specified times, regardless of the presence of malaria parasites at the time of treatment. IPT in pregnancy (IPTp) should be provided to women in areas of sub-Saharan Africa with moderate-to-high malaria transmission. When administered at all routine antenatal visits during the second and third trimesters, IPTp has been shown to reduce the adverse consequences of malaria on maternal and fetal outcomes. IPT in infants (IPTi), delivered through routine vaccination, reduces the incidence of clinical malaria, the prevalence of anaemia, and the frequency of hospital admissions associated with malaria in the first year of life. WHO recommends IPTi as co-administration of SP with DTP2, DTP3 and measles immunization to infants, through routine Expanded Programme on Immunization (EPI) programmes in areas with moderate-to-high malaria transmission and where parasite resistance to SP is not high (prevalence of the pfdhps 540 mutation below 50%).

Seasonal Malaria Chemoprevention (SMC), which involves the administration of monthly courses of antimalarials to children less than five years of age living in areas of highly seasonal malaria transmission, significantly reduces malaria incidence, anaemia and severe morbidity. A complete treatment course of amodiaquine plus sulfadoxine-pyrimethamine (AQ+SP) should be given to children aged between 3 and 59 months at monthly intervals, beginning at the start of the transmission season, to a maximum of four doses during the malaria transmission season (provided both drugs retain sufficient antimalarial efficacy). The target areas for SMC implementation are those where, on average, more than 60% of clinical malaria cases occur within a maximum of 4 months and the clinical attack rate of malaria is greater than 0.1 attack per transmission season in children under five.

Vaccination

An effective vaccine is not yet available for malaria, although several vaccines are under development. A vaccine candidate known as RTS,S/AS01 has shown potentially promising results in clinical trials in Africa. Evidence-based policy recommendations are expected to be made in 2015, and will depend on data to become available in 2014.

Prevention developed for displaced populations

Conventional vector control interventions (e.g. LLINs, IRS) are not always feasible for acute-phase emergencies. The basic, semi-open shelters that often characterize acute-phase emergencies are usually not amenable to effective IRS or hanging LLINs. Space is often an issue and shelters constructed from sharp or abrasive materials very rapidly deteriorate LLINs beyond effective use or lifespan. New vector control tools for acute emergencies are needed that:

- place little or no extra burden on implementing agencies;
- can be stockpiled for the long term;
- require minimal behaviour change among implementers and users.

Displaced populations have specific needs different from stable populations, and insecticide treatment of materials – tents, blankets, sheets, clothing and curtains – may be more acceptable and feasible than conventional interventions. At the time of publication, WHO had not made a formal recommendation on insecticide- treated blankets and plastic sheeting for malaria control. WHO will consider vector control products for policy recommendation on an individual basis as data for those products become available. Current evidence is reviewed below.

Insecticide-treated plastic sheeting (ITPS) – Plastic sheeting is increasingly provided in the early stages of humanitarian emergencies to enable affected communities to construct temporary shelters. ITPS, a laminated polyethylene tarpaulin that is impregnated with a pyrethroid during manufacture, is suitable for constructing such shelters. Like IRS, ITPS is only useful against indoor resting mosquitoes. To be effective, ITPS should represent a high proportion of the surfaces that mosquitoes contact (e.g. all shelter walls, and ideally the ceiling, should be constructed of or lined with ITPS). ITPS also has potential for use as the superstructure of pit latrines as it has documented impact on fly population densities. ITPS effectiveness is likely to depend on local vector resistance to pyrethroids and the ratio of ITPS to other mosquito resting sites. In situations where commercially available

ITPS is used, the duration of effectiveness against malaria and safety under field conditions should be independently assessed and reported to build the evidence base. Since insecticide migrates and accumulates at the surface of ITPS during transportation and when it is stored in packaging, the ITPS should be aired before use.

Entomological trials in refugee settlements in Pakistan showed that insecticide was present on ITPS for at least a year and killed mosquitoes coming into contact with it (Graham et al, 2002). Research to determine ITPS effectiveness during its increasing provision as emergency shelter in Africa remains limited. The results of a recently published trial (Burns et al., 2012) are summarized below (see Box 7.1). Given the limited body of evidence, ITPS cannot replace LLINs or IRS, but can be considered by local experts and authorities for use in situations where tarpaulins will be distributed and LLINs or IRS would be impractical.

Box 7.1 **Evidence for ITPS prevention of malaria in an emergency setting in Sierra Leone**

Evidence that ITPS can provide malaria control in emergencies was obtained during two cluster randomized trials conducted in two refugees camps in Sierra Leone. In the first camp, ITPS or untreated plastic sheeting (UPS) were attached to inner walls and ceilings of shelters. In the second camp, ITPS or UPS were used to line ceilings and roofs only.

In the camp with all inner surfaces lined with plastic sheeting, malaria incidence was 1.63 infections per child under UPS and 0.63 under ITPS, a 61% level of protection. In the camp with ceilings lined, the level of protection was only 15%.

ITPS demonstrated potential to prevent malaria when used at high levels of coverage over all interior surfaces but not when restricted to roofs only. Protection continued for at least a year, after which monitoring stopped. Thus, ITPS proved an appropriate and long-lasting method of malaria control in this emergency setting.

Insecticide-impregnated tents – The inner surface of tents can be sprayed using compression sprayers, either in the same way as house spraying or by laying the tent flat before spraying. Data on the efficacy of this approach also remains relatively limited. Tent spraying has been tried with Afghan and Vietnamese refugees, and is carried out each year among nomadic refugees who migrate annually between the Punjab and the mountainous areas of Pakistan. In the latter case, tent spraying has been shown to provide 60–80% protection against falciparum malaria (Bouma et al., 1996). The most suitable insecticides for tent spraying are permethrin and deltamethrin, in suspension concentrate formulation. The wettable powder formulation used for house spraying is not suitable. Data suggest that pyrethroid insecticide may persist for more than 12 months on double-sheeted tents and up to 6 months on single-sheeted tents (Hewitt et al., 1995). Additionally, a tent of canvas interwoven with insecticide-impregnated polyethylene threads was shown to be highly effective against mosquitoes (Graham et al., 2004).

Insecticide-treated hammock nets – These are sometimes promoted for use by populations who sleep outdoors or in hammocks, such as forest workers in Cambodia (Sochantha et al., 2010). However, data about their effectiveness remains limited, and their role in emergencies is also likely to be limited.

Long-lasting impregnated blankets and topsheets – These have the potential to play an important role in malaria prevention during humanitarian emergencies. Blankets are often included in emergency relief kits, while lightweight blankets or topsheets are widely used in tropical areas. One advantage of blankets and topsheets is they can be used anywhere people sleep (e.g. indoors, outdoors, any type of shelter). The wash resistance of these products is consistent with that of LLINs.

Long-lasting permethrin-treated wash-resistant blankets and topsheets are currently undergoing community randomized trials for malaria control. Permethrin treatment of topsheets, blankets and chaddars (i.e. a covering worn by many Afghan women) has been shown to provide 62% protection against falciparum malaria and 46% protection against vivax malaria compared to a placebo treatment among Afghan refugees (Rowland et al., 1999). Studies have shown permethrin-treated sheets among Kenyan nomads provided personal protection to users (Macintyre et al., 2003), but further research is required to determine whether they can provide community protection at high population coverage levels. The personal protection afforded by insecticide-treated blankets is due mostly to the repellent effect of the permethrin. Permethrin is the preferred insecticide because of its low toxicity, at a target dose of 0.5–1 g/m^2 (Graham et al., 2002).

Clarification of potential role and context is needed before generalized recommendations can be made (e.g. can permethrin--treated topsheets, blankets and chaddars be used as an adjunct to ITPS as they have complementary modes of action and both are distributed in emergencies and do not require technical expertise? What accompanying promotional activities would be needed?).

Permethrin-impregnated clothing – This has been used by the US military for over twenty years and is increasingly popular in the leisure industry (e.g. fishing, hunting). Available toxicology data indicates skin sensitization (e.g. temporary rashes) in hypersensitive individuals, but no significant risk to humans, and permethrin has US Environmental Protection Agency (EPA) approval for use on clothing (Commission on Life Sciences, 1994). Efficacy data of permethrin-impregnated clothing in the prevention of malaria is scant.

Prevention for specific circumstances

Outdoor and evening-biting vectors – In South-east Asia and South America particularly, many vector mosquitoes bite in the evenings or mainly outdoors, potentially limiting the effectiveness of LLINs and IRS. For example, where primary and secondary vector species coexist (e.g. Myanmar), IRS or LLINs may control late-biting vectors (e.g. *An. dirus*) and leave evening-biting secondary vectors free to continue residual malaria transmission. However, even in such settings, LLINs and IRS may remain important vector control tools as the *Anopheles* mosquitoes, which they do affect, may be the most efficient malaria vectors. Aerosol insecticides, mosquito coils, and vaporizing mats are used by householders in these settings, but to date, there is not sufficient evidence of cost-effectiveness to justify encouraging their generalized use or providing them on a broader scale. This evidence is currently being collected.

Topical repellents for application on skin or clothing may have some value during emergencies, particularly as an adjunct to LLINs or IRS. Repellents need to be applied appropriately every evening and are relatively expensive, but may provide a temporary solution provided recipients are disciplined about applying them. In acute-phase emergencies, topical repellents may have logistical advantages over other methods (e.g. speed, size), and a trial among Afghan refugees demonstrated up to 50% protective efficacy against falciparum malaria (Rowland et al., 2004).

Larval control – Used successfully in dengue control, larviciding remains a major undertaking and the evidence for impact, feasibility and cost-effectiveness in malaria control is equivocal. Emergency responses can generate risk through the creation of multiple surface wastewater sites (e.g. installation of leaking emergency water distribution facilities in arid zones). Newly displaced populations may create breeding sites (e.g. through digging borrow pits for house construction), something which is difficult to prevent. Sudden environmental changes can create new breeding sites (e.g.

An. sundaicus was observed breeding more frequently in brackish water after the 2006 Asian tsunami). An appropriate response to human-made localized breeding sites is preventing the cause. Larval control should only be considered where vector breeding sites are limited in number, relatively permanent, and can be easily identified and accessed (few, fixed and findable). In sub-Saharan Africa, such places are most frequently found in urban and peri-urban areas; in rural areas breeding sites tend to be widely dispersed and difficult to find, rendering larviciding a poor vector control strategy (see WHO statement: http://www.who.int/malaria/publications/atoz/larviciding_position_statement/en/). Larval control has not been widely used in humanitarian emergencies, and it is not advised unless there is professional entomological expertise available to provide support in defining the intervention strategy. The list of WHOPES-recommended larvicides is found at http://www.who.int/whopes/en/mosquito_larvicides_Sept_2012.pdf. Table 7.6 lists currently recommended larvicides and their alternatives for use in humanitarian emergencies.

Insecticide zooprophylaxis – Malaria control by application of insecticide to the surfaces of domestic animals has been used in chronic emergencies where vectors bite domestic animals as well as humans (e.g. in south and south-west Asia, where *An. culicifacies* and *An. stephensi* feed on cattle more than on humans). The insecticide – usually deltamethrin – is applied to the hair and skin of cattle, goats and sheep by dipping or sponge. Mosquitoes pick up a lethal dose of insecticide when they attempt to feed on the treated animal.

Zooprophylaxis had a similar effect on malaria to that of IRS, when tested in Afghan refugee settlements – at 20% of the cost as far less insecticide was required (Rowland et al., 2001). Insecticide zooprophylaxis is widely used in Afghan refugee settlements in Pakistan, but only works where vectors are highly zoophilic; trials are being considered against *An. arabiensis* in Africa. Given the limitations of this approach and lack of generalizable data, advice from a medical entomologist is necessary before attempting this method in a new area.

Prevention methods that are not recommended

Scrub removal – There is no evidence that cutting down grass and scrub around houses, to reduce the number of mosquito resting sites, has any impact on malaria transmission; it is therefore not recommended.

Ultra-low volume aerosol spraying and fogging – Whether from a plane, vehicle or motorized sprayer, these methods look impressive. However, there is a risk of usage for promotional rather than malaria control purposes, as there

Table 7.6 **Recommended insecticides for larval control of malaria vectors in humanitarian emergencies**

Insecticide compounds and formulation(s)[a]	Dosage (active ingredient)		
	Class group[b]	General (g/ha)	Container breeding mg/L
Bacillus thuringiensis israelensis strain AM65-52, WG (3000 ITU/mg)	BL	125–750[c]	1.53
Bacillus thuringiensis israelensis strain AM65-52, GR (200 ITU/mg)	BL	5,000–20,000[c]	–
Chlorpyrifos EC	OP	11–25	–
Diflubenzuron DT, GR, WP	BU	25–100	0.02–0.25
Novaluron EC	BU	10–100	0.01–0.05
Pyriproxyfen GR	JH	10–50	0.01
Fenthion EC	OP	22–112	–
Pirimiphos–methyl EC	OP	50–500	1
Temephos EC, GR	OP	56–112	1
Spinosad DT, EC, Gr, SC	SP	20–500	0.1–0.5

[a] DT = tablet for direct application; GR = granule; EC = emulsifiable concentration; WG = water-dispersible granule; WP = wettable powder.
[b] BL = Bacterial Larvicide; BU = Benzoylureas; JH = Juvenile Hormone Mimics; OP = Organophosphates; SP = Spinosyns.
[c] Formulated product.

Notes:
1. Reports of the WHOPES Working Group Meetings (available at http://www.who.int/whopes/recommendations/wgm/en/) and the WHOPES publication *Pesticides and their application for control of vectors and pests of public health importance* (available at http://whqlibdoc.who.int/hq/2006/WHO_CDS_NTD_WHOPES_GCDPP_2006.1_eng.pdf) should be consulted for guidance on use and recommendations;
2. The WHO Guidelines for drinking-water quality (http://www.who.int/water_sanitation_health/dwq/gdwq3rev/en/) provides authoritative guidance and should be consulted for application of insecticides in potable water for mosquito larviciding; and
3. WHO recommendations on the use of pesticides in public health are valid ONLY if linked to WHO specifications for their quality control (available at http://www.who.int/whopes/quality/newspecif/en/).

is little evidence of any impact on malaria. Thus, these are generally not recommended, but can be considered in exceptional circumstances (e.g. outdoor biting and resting mosquitoes). However, such activities would need to be maintained over time to have any impact, and professional entomological expertise is required if consideration is given to deploying this strategy.

Selecting and implementing prevention methods

Once it is determined that malaria prevention is feasible and likely to have impact, the next steps are selecting appropriate prevention methods, planning and organizing implementation, and developing monitoring systems. Thus, necessary decisions include:

- whether to start;
- which tools to use;
- when to change tools to fit changing needs;
- how to transition to country-owned and led vector control;
- when to stop.

Selecting prevention interventions

The main selection criteria and information required for prevention are:

- *Suitability and effectiveness compared to other malaria control methods* – Efficacy is the inherent impact that an intervention can have under ideal or controlled conditions, while effectiveness describes what can be achieved under operational conditions. In emergencies, the principal concern is effectiveness under difficult circumstances. All tools considered here have an evidence base generated from trials conducted in emergency situations. It is also important to remember that emergency situations change over time and need to be reviewed regularly, using relevant data, to ensure selected prevention methods are still appropriate.
- *Feasibility of implementation* – This is affected by contextual issues such as health system logistics; effectiveness of information systems; access to high-risk groups; population movement, behaviour and housing; prior experience of vector control; availability of trained human and material resources; possibilities for supervision and monitoring of activities; existing national policy and government commitment; costs and cost effectiveness; and speed of availability and amount of funding.

Activity selection and planning requires comparison of appropriate and available tools for the different emergency stages (e.g. acute, early recovery), affected populations (e.g. sleeping habits, household construction), and vectors (e.g. night biting, indoor resting). Tables 7.7 and 7.8 list geographical and contextual issues for consideration when selecting optimal prevention activities.

LLINs, and to a lesser extent IRS, are the usual tools of choice to achieve full population coverage. Prevention tools developed for displaced populations should also be considered. Table 7.8 provides a comparison of tools and a list of context-specific characteristics to take into consideration when selecting optimal approaches.

Table 7.7 **Geographical issues to consider in selecting prevention activities**

Epidemiological type	Characteristics of malaria	Possible prevention activities
Savannah or grassland (sub-Saharan Africa, Papua New Guinea)	• Generally present throughout the year • Seasonal increase • Drug resistance, mainly *P. falciparum* • Affects mainly children and pregnant women	• Insecticide-treated mosquito nets • Other types of personal protection • Health promotion
Plains and valleys outside Africa (Central America, China, Indian subcontinent)	• Moderate transmission • Often mainly *P. vivax* • Major seasonal variation • Risk of outbreaks	• Spraying of houses • Insecticide-treated mosquito nets • Other types of personal protection • Livestock treatment • Health promotion
Highland and desert fringe (African and south-east Asian high-lands, Sahel, southern Africa, south-west Pacific)	• High risk of outbreaks • Major seasonal variation • Influenced by agricultural practices • Migration may lead to outbreaks	• Spraying of houses may be considered • Localized house spray-ing for prevention of outbreaks • Insecticide-treated mosquito nets • Health promotion
Urban and peri-urban (Africa, South America, south Asia)	• Highly variable transmission • Immunity of the population variable • Specially adapted urban vectors responsible for outbreaks in south Asia	• Insecticide-treated mosquito nets • Other personal protection • Control of breeding sites by larviciding or environmental planning and management • Spraying of houses in selected areas • Health promotion
Forest and forest fringe (south-east Asia, South America)	• Focal intense transmission • Many risk groups, often occupational	• Insecticide-treated mosquito nets • Other personal protection • Consider siting of dwellings • Health promotion

Table 7.8 **Factors for choosing malaria prevention interventions**

	LLINs	IRS	Hammock nets	Tent spraying	Impregnated blankets	ITPS
Housing type	Not for very basic shelters with sharp or abrasive materials	Not for basic, semi-open shelters with multiple gaps	Not for very basic shelters with sharp or abrasive materials	Appropriate for perennial and seasonal malaria zones	All	Basic emergency shelters
Transmission zone	Appropriate for perennial and seasonal malaria zones	Appropriate for perennial and seasonal malaria zones	Appropriate for perennial and seasonal malaria zones	Appropriate for perennial and seasonal malaria zones	Appropriate for all malaria zones, not for hot climates	
Security	High resale value – can be looted, stolen or sold	Not relevant	No evidence of being looted	Not relevant	High resale value – can be looted, stolen or sold	Some resale value – unlikely to be looted or stolen
Primary vector feeding time	Suitable only for night-biting species	Suitable for both early evening and night-biting species	Suitable only for night-biting species	Suitable for both early evening and night-biting species	Suitable only for night-biting species	Suitable for both early evening and night-biting species
Primary vector resting location (indoor/outdoor)	Not relevant	Suitable only for indoor resting species	Not relevant	Suitable only for indoor resting species	Not relevant	Suitable only for indoor resting species
Displacement versus fixed residence	Transportable	Not transportable	Transportable	Transportable	Transportable	Transportable
Sleeping location	Indoor and outdoor	Mainly indoor	Indoor and outdoor	Mainly indoor	Not relevant	Mainly indoor

	LLINs	IRS	Hammock nets	Tent spraying	Impregnated blankets	ITPS
Compliance/ acceptability	Requires compliance	Requires behavioural change to accept spraying and not to replaster	Requires compliance	Owners need to accept that tents are sprayed	Requires compliance	Requires installation support
Protected individuals	Personal protection – any individual sleeping under the net is protected	No personal protection. Individuals get bitten regardless	Personal protection – any individual sleeping under the net is protected	No personal protection. Individuals get bitten regardless	Mainly individuals sleeping under the blanket	No personal protection. Individuals get bitten regardless
Population coverage	Aim at maximum coverage but could also target vulnerable groups	High coverage necessary for impact on transmission. If coverage with LLINs is high, targeted IRS could be used for insecticide resistance management	Aim at maximum coverage but could also be targeted at vulnerable groups	High coverage necessary for impact on transmission. If coverage with LLINs is high, tent spraying could be targeted for insecticide resistance management	Feasible in progressive way (targeted at vulnerable groups)	High coverage necessary for impact on transmission. Cannot be used to manage insecticide resistance
Logistic burden	High (storage, transportation) but can be integrated with basic commodity distribution	High (requires good skills for insecticide application)	High (storage, transportation) but can be integrated with basic commodity distribution	High (can be integrated in shelter provision needs)	High (can be integrated in shelter provision needs)	High (can be integrated in shelter provision needs)

	LLINs	IRS	Hammock nets	Tent spraying	Impregnated blankets	ITPS
Availability	Ubiquitous	High if vectors are susceptible. Managing insecticide resistance could be a challenge	Ubiquitous	High if vectors are susceptible. Managing insecticide resistance could be a challenge	Low to moderate	Moderate (one supplier only)
Evidence base	Yes	Yes	Yes		Yes in low transmission zones. Ongoing in high transmission zones	Ongoing
Timeframe for replacement or re-treatment	Every 1–2 years depending on wear and tear	Every 2–6 months depending on insecticide and length of transmission	Every 1–2 years (or longer) depending on wear and tear	Every 2–6 months depending on insecticide, length of transmission and material being sprayed	Unknown	Less than 6 months
WHO recommendations	Yes	Yes	Need for more evidence	Need for more evidence	Need for more evidence	Need for more evidence

Source: WHO

Planning malaria prevention activities

Information collected during initial assessment (see Chapter 2) should be used to plan and organize malaria prevention activities. Additional specific information (e.g. about vectors in each site) may be needed to guide organization and monitoring of activities. Table 7.9 shows how information about affected populations, vectors and environment can be used in organizing malaria prevention.

Important steps in setting-up and implementing prevention activities are:

- collecting information (e.g. epidemiological, entomological, demographic, logistic) from international sources and local experts;
- where populations are displaced, choosing resettlement sites away from vector breeding sites whenever possible;
- defining control objectives based on problem severity, available resources and the level of health care in the area;
- choosing control measures;
- collaborating with appropriate agencies to build community acceptance (e.g. sensitizing local authorities and communities) of programme implementation;
- recruiting and training local staff;
- implementing chosen control measures, with entomological advice where feasible and appropriate;
- monitoring and evaluating control measures in terms of coverage (availability and use).

Monitoring prevention activities

Programmes should implement a basic entomological monitoring package of 4–5 essential indicators:

- vector presence and identification (i.e. is there really transmission, and if so, what is the vector?);
- time and place of vector feeding (*Note:* if resources do not exist to determine *in situ* these may be inferred by the species identification. However, in cases where there are multiple members of a species complex, such as *An. gambiae* present, it may be necessary – in addition to morphological identification of the complex – to include molecular determination of the species);
- vector resting habits (see *Note* above);
- insecticide susceptibility status (e.g. this can be determined initially with standard WHO Tube or CDC Bottle assays);
- wall bioassays may also be conducted for IRS programmes where appropriate.

Table 7.9 **Information to guide planning of malaria prevention activities**

Information type	Importance	How information can be used
Environment		
Seasonality of transmission according to rainfall, surface water, temperature	Affects selection and timing of interventions	To determine mosquito control strategies and predict outbreaks
Host		
Population size	Indicates total number of people at risk	To plan the amount of supplies needed
Distribution	Indicates accessibility	To determine the type of malaria control activities
Mobility	Increases possibility of outbreaks	To determine the type of malaria control activities
Types of dwellings and location in relation to breeding sites	Open dwellings are difficult to spray; different types of dwellings need different LLIN designs; proximity of breeding sites increases risk.	To determine the type of malaria control activities
Exposure behaviour	If people are outside during biting times their infection risk is higher.	To recommend when LLIN protection is necessary
Vector		
Species	Different species have different behaviours	To influence control strategy
Preferred breeding sites	Indicates which water bodies are important and whether larval control is feasible	To help determine control methods and communication messages
Resting habits (indoors, outdoors)	IRS and LLINs may be more effective against indoor-resting vectors	To help determine control methods and communication messages
Feeding habits (location, time, host preference)	IRS and LLINs may be more effective against indoor biters and also if people are inside at peak biting times	To help determine control methods and communication messages
Seasonal density	Affects seasonal disease patterns	To help determine content of communication messages and timing of control activities
Insecticide susceptibility status	Affects choice of insecticide	For insecticide procurement

Operational aspects of mosquito biology and behaviour

For the purposes of vector control, it is important to have a basic understanding of:

- how to identify different types of mosquito;
- the life-cycle of mosquitoes;
- mosquito behaviour;
- mosquito insecticide susceptibility.

Types of mosquito

Of approximately 3500 known mosquito species, organized into either Anophelinae or Culicinae sub-families, only 40 species of the genus Anopheles are considered important malaria vectors. Of the 420 Anopheles species currently identified, some can also transmit filariasis and some arboviruses, but these are more commonly transmitted by culicine mosquitoes.

Table 7.10 provides three main ways to distinguish Anopheles from culicine mosquitoes of medical importance, such as Aedes and Culex. Aedes mosquitoes are important vectors of yellow fever and dengue, while some Culex species transmit filariasis and Japanese encephalitis.

Table 7.10 **Three ways to distinguish anopheline from culicine mosquitoes**

Characteristics	Anophelines	Culicines
Resting position	Normally rest at an angle to the surface.	Normally rest with the body parallel to the surface.
Wings	Appear to have black and white spots, because dark and pale wing scales are arranged in blocks.	Do not appear to have black and white spots.
Palps in female mosquitoes	Are the same length as the proboscis.	Are only a third of the length of the proboscis.

Mosquito life-cycle

The four main stages in the mosquito life-cycle are egg, larva, pupa and adult. The first three stages are aquatic, lasting approximately 5–14 days depending on species and ambient temperature. Key differences between anopheline and culicine mosquitoes are also indicated below:

- *Egg.* Mosquitoes lay eggs on water, in batches of 70–200, every few days. Anopheles eggs are about 0.5 mm long and float on the surface of water for 2–3 days until they hatch. Unlike Aedes eggs, they cannot survive if they dry out.

- *Larva.* Each egg hatches into a larva, which swims and feeds. Anopheles larvae float horizontally on the surface of the water, while Aedes and Culex larvae hang down into the water with only their breathing tubes at the surface. In tropical areas, mosquitoes can complete the larval stage in 7 days.
- *Pupa.* The larva develops into a comma-shaped pupa, which does not feed but comes to the water surface to breathe. This stage lasts 2–3 days.
- *Adult.* In most species, the female only produces eggs after feeding on blood. A 2–3-day cycle of feeding, resting/developing eggs, and egg-laying is repeated throughout the adult lifespan of the female mosquito. In favourable conditions, this lifespan averages 10–14 days, although some live longer. The adult female lifespan is crucial in malaria transmission, since it takes at least 10 days for malaria parasites ingested by the mosquito to become infective to humans. Thus, only those infected females who live over 10 days will be able to transmit malaria. Factors affecting adult lifespan include temperature, humidity, natural enemies, and vector control measures. For example, where temperatures exceed 35 °C and the humidity is less than 50%, mosquitoes die much sooner.

Vector control in humanitarian emergencies generally focuses on adult stages, though larval stages may be addressed in specific circumstances. Figure 7.3 shows differences in each stage of the life cycle of anopheline and culicine mosquitoes.

Mosquito behaviour

Aspects of mosquito behaviour that are particularly relevant for vector control are: (i) resting location; (ii) feeding time and location; (iii) host preference; (iv) flight range; and (v) breeding site preferences.

Resting location – After feeding, a female adult mosquito needs to rest while digesting and while developing eggs. Some species rest indoors, while others find outdoor resting sites. It is easier to control indoor-resting vectors with IRS (e.g. IRS is used to control indoor-resting vectors *An. culicifacies* and *An. stephensi* in the Indian subcontinent and *An. gambiae* and *An. funestus* in Africa).

Feeding time and location – Most Anopheles feed at night and each species has a preferred feeding time. For example, *An. farauti* in the Solomon Islands and *An. maculatus* in Myanmar usually feed in the early evening as soon as it gets dark, while *An. gambiae* in Africa and *An. dirus* in South-east Asia usually feed late at night. As with resting locations, some mosquitoes prefer to feed inside houses and others outside. IRS is more effective for

Figure 7.3 **Life-cycles of anopheline and culicine mosquitoes**

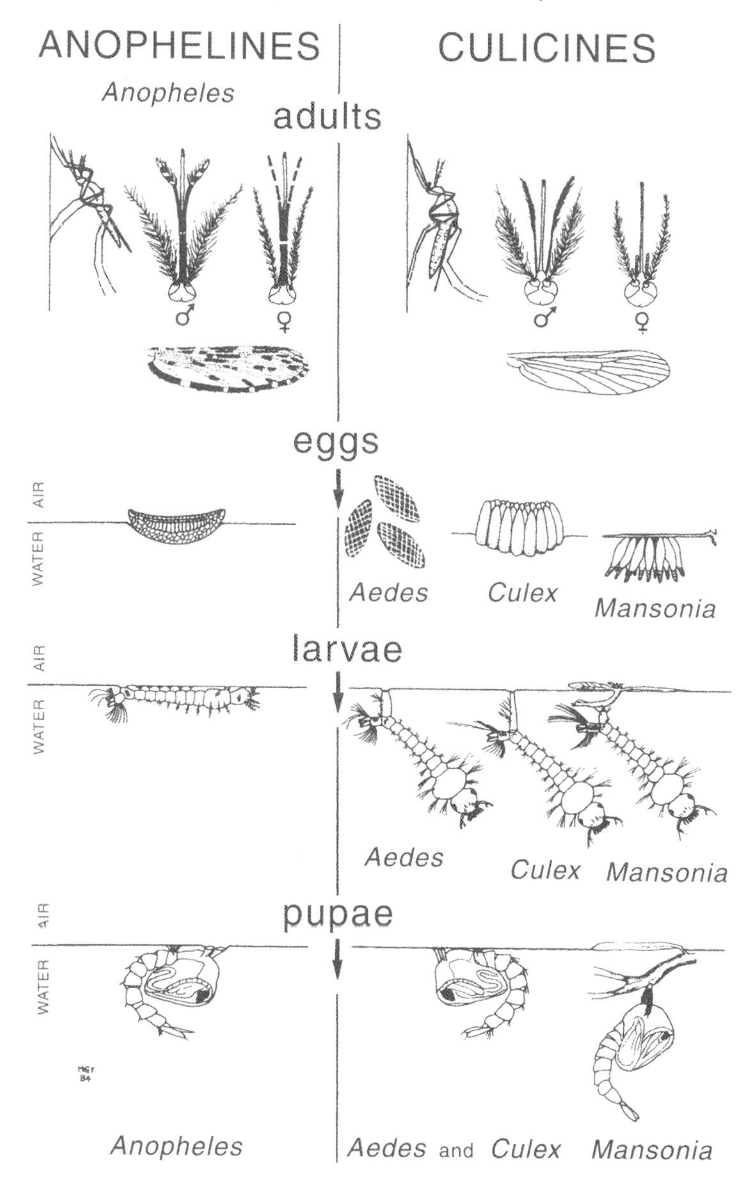

Source: Warrell DA, Gilles HM, eds (2002). *Essential malariology*, 4th ed. London, Hodder Arnold. Reproduced by kind permission of the publisher.

controlling vectors that feed and/or rest indoors while LLINs can be effective in controlling vectors that feed outside.

Host preference – Some anopheline species prefer to feed on humans and others on animals, while several species feed on both. Species with a strong preference for feeding on humans (e.g. *An. gambiae* in Africa) tend to be the most dangerous vectors because they are the most likely to pick up and pass on malaria parasites.

Flight range – Most Anopheles mosquitoes do not fly more than 3 km from the breeding site from which they emerged. As such, malaria risk can be reduced by siting dwellings more than 3 km from known breeding sites.

Preferred breeding sites – Anopheles mosquitoes have different breeding site preferences (i.e. larval habitats), and lay eggs in a wide variety of types and sizes of water bodies, depending on individual species.

A source of vector information and identification keys is the US Armed Forces Pest Management Board, which publishes information on malaria vectors in its disease vector ecology profiles (currently http://www.afpmb.org/content/disease-vector-ecology-profiles). Tables 7.11–7.14 summarize the relevant features of important malaria vectors in Africa, South and South-east Asia, and South America.

Table 7.11 **Important malaria vectors in Africa**

***Anopheles* species**	**Resting location**	**Feeding time and location**	**Host preferences**	**Breeding sites**
An. gambiae	Mainly inside	Mainly late; Inside	Mainly human	Sunlit temporary pools, rice fields.
An. arabiensis	Outside and inside	Mainly late; Outside and inside	Human and animal	Temporary pools, rice fields.
An. funestus	Inside	Mainly late; Inside	Mainly human	Semi-permanent and permanent, especially with vegetation, swamps, slow streams, ditch edges.
An. melas	Outside and inside	Mainly late; Outside and inside	Human and animal	Saltwater lagoons, mangrove swamps.
An. merus	Outside and inside	Mainly late; Outside and inside	Mainly animal	Saltwater lagoons, mangrove swamps.

Source: Mehra (1995)

Table 7.12 **Important malaria vectors in South-east Asia (e.g. Myanmar, Thailand, Cambodia, Lao PDR, Viet Nam)**

***Anopheles* species**	**Resting location**	**Feeding time and location**	**Host preferences**	**Breeding sites**
An. dirus complex (7 sibling species, previously described as *An. balabacensis*)	Mainly outside	Mainly late (20:00–22:00h); Outside and inside	Mainly human	Small shady pools, mainly in forests and forest fringe; stream seepages, footprints, wheel ruts, gem pits, hollow logs, wells.
An. minimus (at least 2 sibling species)	Mainly outside but adaptable	From early evening to early morning; Mainly outside but adaptable	Human and animal	Streams in forest/ forest fringe.
An. maculates complex (8 sibling species)	Mainly outside	All night (peak 19:00–24:00h); Mainly outside	Mainly animal	Sunlit streams, ponds, tanks, riverbed pools.
An. sundaicus (2 sibling species)	Outside and inside	All night (peak 20:00–24:00h); Outside and inside	Human and animal	Brackish water near coast, rock pools, river mouths.

Source: Adapted from Meek (1995).

Table 7.13 **Important malaria vectors in South Asia**

***Anopheles* species**	**Resting location**	**Feeding time and location**	**Host preferences**	**Breeding sites**
An. stephensi	Mainly inside	Late evening and night; Outside and inside	Mainly domestic animals (e.g. cattle)	Urban: domestic water containers, construction sites; Rural: borrow pits, river margins, rice fields.
An. culicifacies (at least 5 sibling species)	Mainly inside	Late evening and night; Outside and inside	Mainly domestic animals (e.g. cattle)	Clean water, river margins, rice fields, pits, pools.

Source: WHO (2005)

Table 7.14 **Important malaria vectors in South America**

***Anopheles* species**	**Resting location**	**Feeding time and location**	**Host preferences**	**Breeding sites**
An. albimanus (Central and northern South America)	Mainly outside	Late evening; Inside	Mainly domestic animals (e.g. cattle, horses), 20% human	Stagnant water, flooded pasture, water with 25% emergent vegetation.
An. darlingi (Mexico to Brazil)	Outside and inside	All night (peaks dusk and dawn); Mainly inside	Human	Temporary freshwater, deforestation and mining areas.
An. aquasalis (coastal, from Venezuela to Brazil)	Outside	Late night; Outside	Mainly domestic animals (e.g. cattle, horses, pigs)	Brackish water, mangrove swamps, temporary freshwater.

Source: WHO (2005)

Insecticide susceptibility

Mosquitoes that become resistant to a particular insecticide are less likely to be killed by that insecticide. A species of mosquito that is exposed to an insecticide for many generations may develop physiological resistance. This is an evolutionary process in which mosquitoes that carry genes for resistance survive exposure to the insecticide and pass on these genes to the next generation. Selection of mosquitoes resistant to DDT was one of the reasons for the failure of malaria eradication efforts in the 1950s and 1960s. The evolution of insecticide resistance does not necessarily mean that the insecticide no longer controls malaria. If only a small proportion of mosquitoes carry the resistance gene or if it is a weak resistance, the insecticide may continue to kill mosquitoes before they become old enough to transmit malaria parasites.

The primary malaria prevention tools, LLINs and IRS, remain highly effective in most settings. However, insecticide resistance among malaria vectors has been identified in 64 countries with ongoing malaria transmission, with resistance detected to all four classes of insecticide used in public health (organochlorines, organophosphates, carbamates, pyrethroids). For example, IRS with pyrethroids showed reduced effectiveness against resistant *An. funestus* in South Africa and *An. gambiae* in Equatorial Guinea (Bioko Island), necessitating a switch to IRS with non-pyrethroids. Evidence from some countries in West Africa indicates that less intact LLINs may be less protective against pyrethroid-resistant *An. gambiae* in selected areas (Asidi et al., 2012).

An important driver of increasing resistance is over-reliance globally on pyrethroids for vector control. Pyrethroids are the most commonly used insecticides in endemic regions as they are relatively inexpensive, long-lasting, effective at both repelling and killing mosquitoes, and safe for humans. All currently-approved LLINs are treated with pyrethroids, and many IRS programmes still rely on them – despite WHO recommendations that the use of pyrethroid IRS be discontinued where LLINs are widely used. Agricultural applications in some areas may also have contributed to resistance. The WHOPES and the Global Malaria Programme websites are a good source of updated information on appropriate insecticides for malaria control and in relation to insecticide resistance management.

Resistance management

All vector control programmes should incorporate insecticide resistance management policies and activities at onset, rather than after resistance has appeared or grown. The Global Plan for Insecticide Resistance Management in Malaria Vectors (GPIRM) (WHO, 2012a) provides a framework for insecticide resistance monitoring and management at global and national levels, for implementation in both stable national programmes and emergency settings – http://www.who.int/malaria/publications/atoz/gpirm/en/index.html Several insecticide resistance management strategies exist for vector control using LLINs and IRS, including:

- rotation of IRS insecticides (e.g. as a minimum strategy, this means regularly alternating between insecticide classes with different modes of action – changing from one pyrethroid to another is not rotation. A pyrethroid may be used as one rotation element, except where there is high LLIN coverage);
- combination interventions (e.g. the combination of LLINs with non-pyrethroid IRS);
- mosaic spraying (e.g. using different insecticides in proximity – like tiles on a mosaic).

The GPIRM recommends pre-emptive rotation for IRS, and case-by-case consideration of focal non-pyrethroid IRS combinations in resistance hotspots where LLINs are used extensively.

Resistance monitoring requires much greater attention and resources today than in the past. Ideally resistance data should be collected from several locations once a year and tracked not only with conventional bioassays, but also with molecular genotyping methods. This requires expertise that is rarely possible in emergencies, but it is desirable to involve national control programmes in data gathering and to share information with them.

In humanitarian programmes where LLINs are to be used, it is very useful to confirm local insecticide susceptibility. Even in countries with high levels of resistance, LLIN usage can provide mechanical personal protection effect. If IRS is to be used, it is mandatory to do susceptibility monitoring. If the insecticide used in IRS is no longer achieving the level of malaria control it previously did, it should be changed. As the alternatives to pyrethroids tend to be more toxic, their application requires better protective materials for spray teams.

Before procuring insecticide, funding agencies should seek advice on local insecticide resistance or take steps to have it measured or monitored by a national control programme or research organization. Updated WHO recommendations on resistance testing methods, and on the collation and interpretation of such data, will be released in early 2013. The impact of resistance on the effectiveness of vector control is a key question. Monitoring schemes can help to show whether vector control operations have less impact in areas with relatively high levels of resistance. Establishment of a global database on resistance is planned by the GPIRM. Until then, control programme data, the African Network on Vector Resistance (ANVR), and other sources should be consulted.

References

- Asidi A et al., (2012). Loss of household protection from use of insecticide-treated nets against pyrethroid-resistant mosquitoes, Benin. *Emerging Infectious Diseases*. 2012 July; 18(7): 1101–1106.
- Bouma M et al. (1996). Malaria control using permethrin applied to tents of nomadic Afghan refugees in northern Pakistan. *Bulletin of the World Health Organization*, 74:413–421.
- Burns M et al., (2012). Insecticide-treated plastic sheeting for emergency malaria prevention and shelter among displaced populations: an observational cohort study in a refugee setting in Sierra Leone. *Am. J. Trop. Med. Hyg*. 2012 August; 87(2): 242–250.
- Commission on Life Sciences (1994). *Health Effects of Permethrin-Impregnated Army Battle-Dress Uniforms*. National Academy Press, Washington DC.

- Graham K et al. (2002). Comparison of three pyrethroid treatments of top-sheets for malaria control in emergencies: entomological and user acceptance studies in an Afghan refugee camp in Pakistan. *Medical and Veterinary Entomology*, 16:199–207.
- Graham K et al. (2004). Tents pre-treated with insecticide for malaria control in refugee camps: an entomological evaluation. *Malaria Journal*, 3: 25.
- Hewitt S et al., (1995). Pyrethroid sprayed tents for malaria control: an entomological evaluation in Pakistan. *Medical and Veterinary Entomology*, 9:344–352.
- Macintyre K et al., (2003). A new tool for malaria prevention? Results of a trial of permethrin-impregnated bedsheets (shukas) in an area of unstable transmission. *International Journal of Epidemiology*, 32:157–160.
- Meek SR (1995). Vector control in some countries of South-east Asia: comparing the vectors and the strategies. *Annals of Tropical Medicine and Parasitology*, 89:135–147.
- Mehra S et al. (1995). *Partnerships for change and communications: guidelines for malaria control*. Geneva, World Health Organization.
- Rowland M et al. (1999). Permethrin-treated chaddars and top-sheets: appropriate technology for protection against malaria in Afghanistan and other complex emergencies. *Transactions of the Royal Society of Tropical Medicine and Hygiene*, 93: 465–472.
- Rowland M et al., (2001). Control of malaria in Pakistan by applying deltamethrin insecticide to cattle: a community-randomised trial. *Lancet*, 357:1837–1841.
- Rowland M et al., (2004). DEET mosquito repellent provides personal protection against malaria: a household randomized trial in an Afghan refugee camp in Pakistan. *Tropical Medicine and International Health*, 9:335–342.
- Sochantha T et al. (2010): Personal protection by long-lasting insecticidal hammocks against the bites of forest malaria vectors. *Tropical Medicine & International Health*, 15:336–341.
- WHO (2005). *Malaria control in complex emergencies: an inter-agency field handbook*. Geneva, World Health Organization. http://www.who.int/malaria/publications/atoz/924159389X/en/
- WHO (2010). *World Malaria Report 2010*. Geneva, World Health Organization. http://www.who.int/malaria/publications/atoz/9789241564106/en/index.html

- WHO (2012). *WHO interim position statement on larviciding in sub-Saharan Africa.* Geneva, World Health Organization. http://www.who.int/malaria/publications/atoz/larviciding_position_statement/en/
- WHO (2012a). *Global Plan for Insecticide Resistance Management in malaria vectors.* Geneva, World Health Organization. http://www.who.int/malaria/publications/atoz/gpirm/en/index.html
- WHO (2013). *Indoor residual spraying: an operational manual for malaria transmission, control and elimination.* Geneva, World Health Organization. http://www.who.int/malaria/publications/atoz/9789241505123/en/index.html

Finding out more

- Corbel V et al., (2004). Dosage-dependent effects of permethrin-treated nets on the behaviour of *Anopheles gambiae* and the selection of pyrethroid resistance.
- USAID (2010). *Presidents Malaria Initiative BMP Manual: Best Management Practices (BMP) for Indoor Residual Spraying (IRS) in Vector Control Interventions*, USAID, CDC. http://pmi.gov/technical/pest/bmp_manual_aug10.pdf
- WHO/GMP website: www.who.int/malaria
- WHOPES website: http://www.who.int/whopes/en/

CHAPTER 8
Community participation

This chapter:

- outlines principles and key messages for community participation and health communication related to malaria control in humanitarian emergencies
- describes how to involve communities in malaria control and health education
- identifies issues to consider when planning and implementing health communications related to malaria control.

Principles for community participation

Background

Community participation and health education are often seen as being of low priority in humanitarian emergencies. In conflict-affected areas, reaching populations – especially IDPs – poses multiple challenges, including access to those displaced and security threats. Despite this, community participation and health education are essential to the success of malaria control interventions in emergencies (see Table 8.1). Health education has been defined by WHO as "...any combination of learning experiences designed to help individuals and communities improve their health, by increasing their knowledge or influencing their attitudes." Health education can be conducted at any point during an emergency, directed at either individuals or communities. Communities provide front-line protection from malaria during an emergency because:

- local knowledge of local risks ensures that the actual needs of the community are addressed;
- local actions prevent risks at the source, by avoiding exposure to local hazards;
- a prepared, active and well-organized community can reduce both risks and impact of emergencies;

- many lives can be saved in the first hours after an emergency before external help arrives (Global Health Workforce Alliance 2011).

A well-prepared community can prevent excess illness and death from malaria when an emergency occurs and can prevent outbreaks of malaria when detection, prevention, and treatment systems have been disrupted or populations are mobile. Successful prevention and preparedness requires the active involvement of communities in malaria control programmes and that behaviour and social-change programmes are in place before an emergency occurs. After the onset of an emergency, key messages and risk communication strategies need to be ready for rapid implementation.

Guiding principles for effective education participation and malaria-related health communication during emergencies are to: (i) establish mechanisms that promote good communication and enable representatives of the displaced or affected population, the host population, and local and international emergency partners to participate in malaria control; (ii) implement key messages that stress preventive strategies, how to identify malaria, and need for prompt diagnostic testing, complete treatment, and supportive care.

Key messages

Work with existing messages and communication strategies – Education targeting malaria-specific risk communication and behaviour and social change should be integrated into an overall health communications strategy developed and adapted to local settings as part of regular malaria control efforts, particularly in malaria-endemic or epidemic-prone areas. It is important not to disrupt this system during an emergency but to continue and extend it to the affected areas. The messages below are examples of key messages during any emergency.

Key message 1: Know the disease

- Malaria is transmitted through the bites of some mosquitoes;
- Malaria frequently causes fever;
- Wherever malaria is present, children and pregnant women are in particular danger;
- In an emergency, malaria prevention and treatment programmes can be disrupted and families and communities need to know what to do.

Rationale: Community members, especially caregivers, should be informed about malaria (e.g. how it is transmitted, symptoms of the disease, how to prevent it, what to do if someone becomes ill with fever). Communities and families have a right to information about potential changes in programmes during emergencies, about how to continue to access care and prevention, and about how to prepare for increased risks. Intended behaviour outcomes include:

- consulting with community health workers (CHWs) on basic information about malaria and how it can affect family members, especially children and pregnant women;
- consulting with CHWs about how the family can prepare for possible emergencies, including how to access care during malaria outbreaks and what the role of family members may be.

Key message 2: Recognize danger signs and seek prompt care

Rationale: People in the community, especially caregivers, should know the danger signs of malaria. It is important to be able to identify children and other community members who require immediate medical attention and need to be taken to health facilities. During epidemics, all persons regardless of age are often at risk, especially in settings where malaria was previously absent or infrequent. However, children and pregnant women should be prioritized for care in most settings as they are likely to suffer adverse outcomes from malaria more quickly. Intended behaviour outcomes include:

- families learn malaria danger signs from the health care worker;
- families take the sick family member to a health care provider for immediate care and follow the advice of the health worker;
- families prioritize care seeking for children and pregnant women.

Key message 3: Malaria is preventable

Rationale: Sleeping under an LLIN is the best way to prevent mosquito bites during the night. Since children and pregnant women are at most risk of severe illness and death from malaria, they should be given priority in sleeping under LLINs if there are not enough LLINs for everyone. In moderate-to-high-transmission areas in Africa, pregnant women can help protect themselves by taking IPTp. Intended behaviour outcomes include:

- everyone should sleep under an insecticide-treated bednet, preferably an LLIN every night – even when they don't see or hear mosquitoes around;
- if there are not enough LLINs, then preference should be given to children and pregnant women;
- pregnant women should take IPTp, where recommended by health workers, and sleep under an LLIN.

Key message 4: Malaria is treatable and you can learn to care for family members sick with fever

Rationale: A child suffering or recovering from malaria needs plenty of liquids and food. Preventing dehydration and ensuring continued and adequate nutrition are key practices that contribute to limiting severe complications and death from common childhood illnesses such as malaria. When a child is sick, families should be encouraged to feed and offer additional fluids. This should include continued and increased breastfeeding.

A child with a fever should be examined immediately by a trained health worker and receive appropriate antimalarial treatment as soon as possible if diagnosed with malaria. Intended behaviour outcomes include:

- mothers continue breastfeeding sick babies and increase frequency or duration of feedings;
- families continue to feed the sick person and provide additional fluids for non-nursing children, such as clean water, soups, broths, teas, fruit juices.

Involving communities

Community involvement and effective health education and communication play a critical role in both treatment-seeking and malaria prevention. Table 8.1 provides some benefits of community participation.

Table 8.1 **Community participation and health education in malaria control**

Malaria treatment-seeking
Community understanding and action influence the effectiveness of case management. One of the reasons for high malaria mortality in both emergency and non-emergency situations is late presentation of cases to, or failure to seek treatment from, health facilities. Communities need to be able to recognize the signs of malaria illness and seek diagnostic testing and effective treatment promptly.
Malaria prevention
Community understanding and action also influence the effectiveness of most prevention methods. For example, communities need to know why the use of LLINs is important and why young children and pregnant women are particularly vulnerable and should be prioritized to sleep under LLINs when adequate supplies for universal coverage are not available. Community cooperation is essential for the success of IRS.

Identifying and coordinating with partners

During an emergency, it is important to coordinate with other partners working on community-based programmes and communication strategies.

Although malaria is a specific topic, education messages and strategies will be rolled out through inclusive, multi-disciplinary social mobilization programmes and other channels such as the media. Key messages for malaria and mobilization of communities should be coordinated through a specific social mobilization working group under the overall health sector/cluster coordination mechanism. This group will help define the messages and the strategies for dissemination.

It is essential to remember that affected communities are vital partners to include when identifying priority needs and interventions to address those needs. Communication with the affected communities assures that priorities are jointly set and helps the affected population feel more in control of their situation. Not including affected communities as partners risks malaria-control interventions not being supported or not being recognized as important. Community involvement ensures that malaria control activities and communication strategies are socially and culturally appropriate.

Messages can be brought to communities through a variety of interpersonal and media mechanisms used in risk communications, aimed at rapid dissemination of urgent information, and longer-term behaviour change communications strategies. The International Federation of Red Cross and Red Crescent Societies (IFRC) and UNICEF are international agencies with particular expertise in risk and behaviour change communications during emergencies.

Risk communication

Risk communication is an interactive process of exchanging information and perceptions on risk among assessors/evaluators, implementers, and other interested parties (e.g. affected communities) and is an integral part of risk assessment. It can be used in malaria control in emergencies, for example, to inform communities of outbreak risks, of where to seek treatment if routine services have been disrupted, or of how to report a problem that they have observed.

Behaviour change communication (BCC)

BCC is most commonly used during the post-acute stabilization phase and during chronic emergencies. It is an approach defined as the strategic use of communication to promote positive health outcomes, based on established theoretical models of behaviour change. BCC employs a systematic process beginning with formative research and behaviour analysis, communication planning, implementation, monitoring, and evaluation. Key features are:

- inclusion of both children and adults;
- sharing of relevant and action-oriented information;
- using a consultative process among technical experts, community members, local change agents, and communication specialists (UNICEF-ROSA, 2006).

BCC can be used in malaria control, for example, to encourage appropriate use of LLINs, compliance with antimalarial drug regimens, or increase participation in IRS campaigns (Mushi et al, 2008; Panter-Brick, et al, 2006). Information, education, and communication (IEC), which focuses on knowledge change, can be combined with more participatory BCC to enhance communication with affected communities.

Two recent projects that have used BCC to increase involvement in malaria control activities in the context of emergency settings are (i) the *Health Initiative in the Private Sector Project* (HIPS) working with IDPs in northern Uganda, and (ii) the *Liberia Rebuilding Basic Health Services* (RBHS) Project. Applying lessons learnt from such projects that have used BCC in malaria control will increase the likelihood of successful implementation. These lessons include:

- reinforce messages multiple times with both interpersonal and community support;
- obtain financial and political support;
- focus on positive outcome expectations that are relevant to the target audience;
- provide messages in simple formats that are repeated in many different ways and contexts;
- address situational constraints that might affect people's abilities to change;
- develop new skills in the targeted audience (Malaria Consortium, 2009; Panter-Brick et al, 2006).

The first example in Box 8.1 illustrates the importance of giving community leaders an opportunity to express their concerns and be understood.

Box 8.2 offers an example of adapting symptom control measures to local belief systems to improve their acceptability.

Understanding communities

To strengthen communication with affected communities, develop appropriate communication strategies, and increase positive behaviour, it is necessary to understand what might influence behaviours. In emergency

Box 8.1 **Importance of communicating and listening to community concerns**

In an African refugee camp, antimalarial drug efficacy studies were planned for children under 5 years of age. To ensure that this was acceptable to the community, meetings were held with camp officials, CHWs and refugee leaders prior to initiating the drug studies. One of the major concerns voiced by the community was why the studies only included children younger than 5 years of age when adults were also suffering from malaria. Once it was clarified that children are particularly vulnerable to malaria and that appropriate treatment for children would also benefit the adult population, community leaders agreed to support the research, and messages were sent out to the residential blocks encouraging people to participate if asked.

Box 8.2 **Importance of understanding local context**

Refugees from Angola have been settling in the north-west border area of Zambia since the late 1960s. Because both the displaced and host populations believed that people with fever should not drink water, promotion of hydration in people with febrile illness, particularly children, was failing. However, when food colouring and sugar were added to the water and the resulting solution was described as medicinal, hydration with this solution became acceptable.

contexts, 'community' may involve those displaced, host communities, and the humanitarian community as these all interact with one another in meeting health needs.

Social factors (e.g. structures, organization) and cultural factors (e.g. language, beliefs, common behaviours) play an important role in malaria control during emergencies. To promote effective community participation it is helpful to learn about social and cultural contexts within displaced and host communities. Displaced populations may not share the same culture as host communities. Within displaced communities, there may be differing ethnicities, values, and priorities represented. Within the humanitarian community, different organizations – UN, bilateral and government agencies, international and local NGOs – may share many values and beliefs but will also have different organizational cultures and priorities. Successful coordination depends on recognizing and working with these differences.

For both displaced and host populations, it is essential to identify how social organization has been changed by the emergency and how social structures similar to those pre-existing the emergency can redevelop as quickly as possible. It is particularly important to:

- ask about social organization, the roles played by different groups and individuals, and who influences attitudes and behaviours. Learning how residential units are organized will provide information about the social structure and any political divisions that might affect implementation of control activities;
- be sensitive to people's need to regain control, self-determination, and a sense of meaning in their lives.
- identify all vulnerable groups – not only those that are physiologically vulnerable (e.g. pregnant women and young children) but those that may be vulnerable for political, ethnic, social or cultural reasons. For example, minority ethnic groups may be politically vulnerable and so unable to access health facilities or to participate fully in community education or prevention activities.

Table 8.2 **Potential roles of community organizations after crisis**

Type of organization	Example organizations and networks	Potential role after crisis/displacement
Formal	Government services (e.g. health, welfare, education, justice)	May be absent, run by non-state (NGO, opposition) agencies, and/or focused on religious or other groups.
	UN and donor support	Will change over time, with different agencies leading (e.g. cluster approach).
Semi-formal	Religious organizations	Religious leaders and organizations may be significantly involved throughout all phases of a crisis.
	Political organizations	Political organizations may focus primarily on national influence or become more active in later phases.
	Opinion leaders	These may be formal or informal; community health workers or midwives may play this role.
Informal/Social	Family networks	May be absent or widely dispersed and need to be rebuilt.
	Social networks	Existing networks may be strengthened or weakened; ad-hoc networks may form through shared displacement or other experiences.

Source: Adapted by N Howard from Person & Cotton (1996)

Working with communities to identify priority needs and activities

Involving communities in defining priority needs and activities (see Table 8.3) can improve the effectiveness of malaria control activities. Malaria control priorities should be based on community needs and perspectives, but will also depend on resources available to agencies and on what can be achieved. Given the limited resources often available in emergencies, it is important to think creatively with the community to determine the best ways to link malaria control activities with other public health interventions. For example, UNICEF has implemented LLIN distribution during routine ANC visits, child immunization campaigns, and child health days. In the acute phase of an emergency, however, health professionals often determine immediate needs (e.g. dealing with a malaria outbreak). Once the situation stabilizes, community priorities and activities (e.g. prevention) can receive more attention.

Table 8.3 **Involving the community**

- Find out how to show respect to community leaders.
- Listen to the concerns of displaced and host populations.
- Allow community representatives to define their own concerns and priorities first. This is important because people in some cultures will agree with those they perceive to be more powerful, even if they do not support what is being said. As such, community agreement to proposals from international agencies may not reflect the community's actual concerns and priorities.
- Define together the problems, causes and possible solutions.
- Involve community representatives in planning and implementing qualitative research to gather information and to analyse the data collected.
- Relate malaria control activities to local ideas about cause, symptoms and cure. If feasible, involve traditional healers in case management and referral.
- Provide clear information about how and where malaria control activities will take place (e.g. specify the location of clinics and outreach centres and the timing and location of spraying and LLIN distribution). If possible, involve community representatives in conducting prevention measures such as house spraying (IRS) and LLIN distribution.
- Be sensitive to gender and ethnicity. Community leaders may not represent the views of women, the most vulnerable, or minority groups. Try to include representatives from these groups (e.g. link malaria control to the activities of women's groups or programmes that target women).

Involving communities in information collection

Individuals recruited from the displaced population can help collect information about the community and priorities. As members of the affected population, they can facilitate communication and provide useful detail (e.g. about community leadership, structure, politics). Initial contact will often be with the most educated and influential people, and additional information is needed from other subgroups that may not be represented by community leaders.

One way to learn about the community and identify vulnerable subgroups is to walk through the resettlement areas with community members.

- Walk with trusted volunteers identified by CHWs. If you walk with the most visible or important leaders, community members may be reluctant to voice concerns in front of these leaders or to reveal what is happening in their residences.
- Conduct walks at randomly chosen times of day, on several different days and in different areas.
- Make visits a surprise so that people will be engaged in typical daily activities rather than preparing for a "guest." However, if an unannounced visit violates cultural ideas of politeness or poses a security risk, community leaders should be alerted about the timing and purpose of visits.
- Hold informal conversations with community members during the walk. Ask them about the community. Who is in need of what services? Who has been sick recently? Who is frequently sick? This approach will help to provide information about community dynamics, which some leaders may be reluctant to discuss.
- If possible, take notes as you walk, recording what you observe as well as what people tell you.

To identify priorities, it is essential to collect community answers to the following questions:

- What is the community's previous experience with malaria control activities?
- What malaria control activities are in place?
- What do local people think and do about febrile illnesses, including malaria?
- What local words and terms are used to describe malaria?
- What does the community believe about the causes, symptoms and treatment of malaria?
- What sources of information are used and trusted by the community?

Table 8.4 **Qualitative methods**

Individual interviews
These are guided by a mixture of structured and open-ended questions to explore topics related to malaria control. Interviewees can either be selected purposively (e.g. as key informants, community leaders or health workers) or selected randomly. An advantage of individual interviews is that questions can be flexibly structured and topics raised by the interviewee can be pursued.
Focus group discussions (FGDs)
These involve discussions with groups of 8–15 people with similar backgrounds. A facilitator guides the discussion using a checklist of topics. A second person records what participants say and notes how they interact. FGDs are a useful initial method for determining what topics are important to people and can also be used to clarify or check information collected during individual interviews.
Free listing
Participants are asked to list as many different aspects as possible of the topic of interest e.g. "What methods are used in this community to prevent malaria?" This can generate a wealth of information that can be used to develop questions for interviews, topics for FGDs, and guide the design of malaria control interventions. An advantage of free listing is that it provides a large amount of information in a short time period and can be conducted as a game.
Pile sorting
Individually or in groups, participants are asked to sort the issues they have listed into categories (e.g. asking them to sort a list of malaria treatments into logical categories may provide information about local remedies for different types of malaria). Thus the information identified through free listing can be used to find out how people perceive and make sense of various issues, which can feed into the design of interventions.
Mapping
Uses maps drawn by community members to show different community features, such as where clinics are located or where malaria is a particular problem. Mapping provides useful information about community perspectives.
Venn diagrams
These are developed in participation with community members to illustrate how groups and individuals in the community interact with each other (e.g. degree of cooperation or influence, potential roles). Diagrams are developed by listing community groups and key individuals and drawing different sized circles to represent each, with circle size indicating decision-making importance, distance between circles indicating degree of contact, and overlap indicating level of interaction (e.g. large overlap for high interaction). This can help identify who might support or oppose malaria control activities.

- When and where do people seek treatment for malaria?
- What does the community do to prevent malaria?
- What are the community's priorities for malaria control?
- What factors would encourage the community to take prevention measures and to seek treatment?
- What factors might stop the community from taking prevention measures or seeking treatment?

This information can be collected qualitatively. Table 8.4 summarizes different qualitative methods. Ideally, a social scientist should help design qualitative research tools and collect information. Local expertise can also be consulted. If time and resources are limited, rapid assessment techniques can be used to collect information. Annex IX includes examples of questions to guide development of malaria control activities.

Using qualitative research approaches to working with and learning from communities can be daunting for humanitarian workers who have not previously used these methods. Table 8.5 offers reflections that may help in developing a rapport with communities.

Involving the community in implementation

The affected community is a partner, not the target, during control activities. Individuals who are known and trusted in the community must be involved in implementation. Consider how a cadre of community "activists" can be developed and how community organizations (e.g. women's groups, centres of worship) can be used to mobilize others to participate in malaria control activities.

A joint agency statement on the importance of scaling up a community-based health workforce for emergencies, released in October 2011 and endorsed by the humanitarian community (Global Health Workforce Alliance et al, 2011), illustrates the critical role of community involvement and outlines the roles of CHWs in all aspects of public health, including malaria control. Community members and CHWs often already play an important role in a range of malaria control activities, working to detect malaria with RDTs, providing treatment, identifying danger signs and referring severe cases to health centres, as well as delivering LLINs and health communication messages. Opportunities for case management include:

- integrating traditional healers into some health clinics (see Box 8.3);
- training community members as peer role models to improve the process of giving antimalarials to small children (see Box 8.4).

Table 8.5 **Reflections on 'lessons learnt' from fieldwork**

Do your homework before going to the field (e.g. relevant terms and acronyms, political/administrative divisions). Once you are on the ground, you will receive a lot of information quickly. The more you understand during data collection, the easier it will be for you to ask relevant follow-up questions.
Choose key informants wisely. Political leaders may not be the most knowledgeable or represent all segments of society (e.g. district-level officials may be politically appointed, may not be local, and may not understand the nuances of the communities that they serve).
Learn what is and what is not culturally appropriate, and act accordingly. Your behaviour should reflect local culture, including clothes, eating, greeting, body language, and the way in which you sit during an interview (e.g. level relative to others, not showing the soles of your shoes). Greet people with respect and humility. If possible, learn greetings and words of respect in the local language. These practices are not only polite, they may also help community members feel more comfortable sharing their thoughts with you.
Be sensitive in your use of technology. Use of items such as laptops, tablets, and smartphones can distance you from communities who are not accustomed to them. It may also attract unwanted attention, frighten people, and compromise your safety in the field.
Expect the unexpected, and make the best of unplanned disruptions to field work. Even during displacement, community members may be busy attending to tasks of daily living, such as finding firewood. It can be frustrating when you are scheduled to meet with the community and few or no people show up. Rather than being frustrated, use the unexpected free time to speak with your local staff or community members who are present, to plan additional work, or simply to rest.
Keep eyes and ears open, even when your notebook is closed. Some of your most profound insights about the community of interest will occur at unexpected moments – with a driver in transit to a research site, with staff over lunch, or with a fruit vendor after work.

Source: adapted from Becknell (2012).

Box 8.3 **Involving traditional healers in case management in Cambodia**

In the paediatric ward of a large refugee camp on the Cambodia-Thailand border, traditional healers (Kru Khmer) were encouraged to work alongside health workers. Ward rounds included both clinical staff and Kru Khmer. Families liked this approach and the Kru Khmer provided valuable inputs into clinical decisions. Together, clinical staff and the Kru Khmer were able to provide complementary care, rather than competing treatment options. For example, a child might receive standard treatment for malaria along with a herbal drink or massage from a traditional healer. The Kru Khmer appreciated the recognition of their role and the health workers appreciated the cultural knowledge of the traditional healers.

Box 8.4 **Caregivers as peer role models**

One of the unexpected benefits of a drug efficacy trial in a refugee camp in the United Republic of Tanzania was the emergence of caregivers as peer role models for the rest of the community. Those caring for children enrolled in the trial learnt how administer antimalarial drugs properly. They were able to watch the team members administer the drugs used in the trial and ask questions about what they were doing. This provided an excellent opportunity to teach caregivers – mothers, fathers, older siblings and grandparents – safer techniques for holding children while giving medication, different ways to give the drugs (e.g. crushing tablets to mix with liquid in a bottle or with a spoon), and what to do if a child vomited the drug shortly after administration. Caregivers said that, because regular health workers were often too busy to offer these explanations, they had not always known what to do with a child who refused to take medication or had vomited. Now that they knew what to do and they planned to share this knowledge with their family, friends, and neighbours.

Opportunities for prevention include:

- training community members as community health educators (see Box 8.5);
- employing community members to distribute or repair LLINs, or to serve on IRS spray teams.

Box 8.5 **Training community health educators in malaria prevention in refugee camps on the Myanmar–Thailand border**

NGOs trained members of local ethnic minority groups as community health educators in refugee camps along the Myanmar–Thailand border. During the mid-1990s, previously stable camp populations were relocated due to shelling and border incursions. Community health educators ensured the continuity of malaria control activities. As the population moved to a new site, educators were able to go from tent to tent and provide essential services including: (a) information about the new location of health services and general health messages; (b) active case detection for presumed cases of malaria and other common illnesses; and (c) education in the correct use of insecticide-treated nets. Because they belonged to the same ethnic group as the displaced population, community educators were able to bridge the gap between refugees and the agencies working in malaria control. This was particularly important for some of the more isolated ethnic groups who had little previous exposure to, or knowledge of, this kind of health care.

Opportunities for research include training local health workers to interpret the routine information they collect to help in designing malaria control interventions (see Box 8.6).

Box 8.6 **Training community health workers**

In settlements along the Myanmar-Thailand border, NGOs trained community health workers to interpret the routine data that they were collecting, and to increase their understanding of how this information could be used to design malaria control activities. This method, called community-oriented primary care, combined elements of primary health care, public health and community mobilization. Meetings and workshops were held on a regular basis to provide training for CHWs and to review data collection and analysis, implementation, monitoring and evaluation.

Many community approaches will have originally been used in stable situations, and adjustments to commonly applied interventions may be necessary, depending on the current level of security, the relationship between displaced and host communities, and the length of stay in the host community. To improve the success of these interventions it is important to stress the following:

- conduct thorough and repeated trainings;
- institute field supervision;
- use multiple methods to stress key points;
- ensure adequate supplies for interventions (e.g. reliable antimalarial supplies);
- understand the context in which interventions are applied and how that context might affect acceptance of the intervention (Atkinson, et al., 2011).

There is limited evidence to support the assumption that community participation contributes to reductions in disease transmission (Atkinson, 2011). Recognizing this, operational research on common community-based malaria control interventions in emergency settings is needed to determine whether these efforts affect disease transmission in the context of displacement.

Health communication in malaria control

Establishing priorities

The goal of health communication (e.g. risk communication, health education, BCC) should be to inform populations at risk about behaviour changes that could reduce their risk of malaria infection and improve their management of malaria illness. Health communication should address specifically the concerns identified by displaced and host communities, in addition to:

- causes of malaria and the role of mosquitoes in transmission;
- groups that are particularly vulnerable to malaria;
- differences between uncomplicated and severe malaria;
- diagnosis and treatment of malaria, including the importance of diagnostic testing;
- location and opening hours of health services;
- danger signs for severe malaria that mean a patient should go back to the clinic;
- measures that can be taken to protect against mosquito bites, including use of LLINs and other forms of personal protection.

The emphasis placed on each of these points will depend on the phase of the emergency. In the first few days of the acute phase, it is essential that people know both how to recognize the signs of severe malaria and where and how to access care. As an emergency evolves and the situation stabilizes, more attention will be given to preventing transmission or sustaining preventive measures.

Emphasis will also depend on the local context. For example, if the displaced population has moved from a non-endemic area to an endemic area, information about transmission risk will be important.

Designing health communication

In the acute phase of an emergency, with competing urgent priorities, it is seldom possible to develop a formal health communications strategy. During this phase, information passed by word of mouth can be a useful way to publicize urgent messages in the community.

In the post-acute phase as the situation stabilizes, a health communications strategy can be designed and implemented. It is important to invest time and resources in doing this with communities, who should be involved in designing and pretesting messages and making necessary revisions. Failing to do this can waste limited time and resources, producing ineffective messages and materials.

Health communication should focus on **learning** (not teaching), **active participation**, **action**, and **behaviour change.** The principles of successful health communication include:

- *Define objectives* – for example, the desired improvement in knowledge, change in behaviour, or change in attitude;
- *Identify the target audience* – for example, pregnant women, parents of young children, community leaders, traditional healers, drug sellers;
- *Develop clear messages* – in such a way that the target audience can understand them;
- *Provide information about what people can do* – for example, where they can get LLINs or when to seek treatment – with a limited number of clear, simple messages, such as "prompt treatment for fever saves lives", repeated in different ways;
- *Use methods that are culturally acceptable* – and appropriate for the target audience;
- *Deliver messages through trusted and respected individuals and channels*;
- *Provide training and materials* – to those who will conduct health communication (e.g. community educators, health workers, traditional healers).

Effective health communication depends, in part, on understanding aspects of the community being targeted. Specific information from the targeted community, such as knowing where and how people expect to receive important information, will help to inform decisions about the design of communication materials (see Table 8.6).

Visual messages should be clear enough that anyone, including children and people who speak other languages, can understand what they are trying to convey. Written messages should be short, simple, fun, and written in a way that catches the interest of the intended audience. The primary communication objective should be identified and the message focused on communicating that idea. The essence of the message should be delivered through a 'SOCO,' or single, over-riding communication objective.

Implementing health communication

During emergencies, it is important to plan the timing of health communication activities carefully:

- Avoid days or times when other activities are taking place (e.g. rations are being distributed, cultural or religious events);
- Take advantage of occasions when people are already meeting together, for example, meetings with block or camp leaders;

Table 8.6 **Communication channels, messages and methods**

Channels

Effective delivery of health communication messages depends on choosing communication channels appropriate to the community's culture, knowledge, beliefs, values, and literacy level. In some cultures, community leaders and elders may be the best people to transmit messages, while in others, teachers, health workers or the media may have more credibility. Messages can be relayed through community workers, such as those who dig latrines or do manual tasks. Other possible dissemination channels include:

- camp management meetings;
- health and water coordination meetings;
- health facilities;
- community health programmes, such as vaccination campaigns;
- central message boards, particularly in transit areas.

Messages

The following principles should govern the development of messages:

- target the message to the lowest level of literacy;
- explain what people are expected to do and the expected outcomes from their behaviours;
- ensure that messages are realistic and feasible;
- understand local behaviours;
- use local language for the messages;
- ensure that messages are culturally acceptable;
- change messages as the situation changes;
- use creative approaches in the messages: for example, promoting LLINs may be more successful if the messages describe the nets as a way of "preventing nuisance insects from disturbing your sleep", as well as a way to prevent malaria.

Methods

Methods must be appropriate for the setting. Printed messages are fine for literate populations, but oral and visual methods will be better for conveying messages to mixed-literacy populations. In most situations, it is useful to use a mixture of methods. Possible methods include:

- **Print** – posters, flipcharts and picture guides can combine words and pictures. Posters should be displayed in places where people will see them (e.g. clinics, registration centres). Flipcharts and picture guides are useful for face-to-face teaching (e.g. about how to give an antimalarial drug to a young child).
- **Oral** – talks, person-to-person communication, songs, poetry and informal conversations (e.g. to religious or women's groups or at times when people are waiting at the clinic).
- **Visual** – drama, role-play, dance, demonstrations. Drama and similar methods use stories to get across messages. Demonstrations are a useful way to show people how to do something, for example how to use a treated net.
- **Broadcast media** – public broadcasting or radio. Public announcements through loudspeakers or megaphones can be especially useful in the acute phase of an emergency when people are still arriving.

- Alert the community in advance about campaigns or activities (e.g. IRS or distribution of LLINs during immunization campaigns);
- Use opportunities to target specific groups (e.g. pregnant women, mothers of young children). In many emergencies, women spend time waiting in outpatient clinics or at supplementary feeding centres. This time can be used for health communication about malaria (e.g. what to do when a child has fever). Men also bring sick children to clinics and should not be ignored.

Monitoring and evaluation

As with all aspects of malaria control, community participation and health communication activities must be monitored and evaluated to determine whether activities have succeeded in changing targeted behaviour. For example, it may be necessary to assess whether there have been changes in use of individual protection methods – by monitoring and evaluating LLIN coverage and proper usage. Alternatively, it may be important to assess changes in treatment-seeking and case management – by monitoring and evaluating the number of children brought for treatment, treatment compliance, the length of time between initial symptoms and seeking care at a health care facility, or the number of return visits.

References

- Atkinson JA, et al. (2011). The architecture and effect of participation: a systematic review of community participation for communicable disease control and elimination. Implications for malaria control. *Malaria Journal*, 10: 225
- Becknell K (2012). *"Notes from the field: qualitative research methods in Nepal and Sri Lanka."* Lecture given in 'Epidemiologic Methods in Humanitarian Emergencies' course (GH 510), Emory University, Atlanta, 12 Jan 2012.
- Global Health Workforce Alliance (GHWA), WHO, IFRC, UNICEF, UNHCR (2011). *Joint Statement: Scaling up the community-based health workforce for emergencies.* http://www.who.int/workforcealliance/knowledge/publications/alliance/jointstatement_chwemergency_en.pdf
- Malaria Consortium (2009). Summary from the technical workshop on cross-border IEC/behaviour change communication strategies to contain artemisinin resistant malaria. In the context of Bill & Melinda Gates supported project: "A Strategy for the Containment of Artemisinin resistant Malaria Parasites in South-east Asia." Meeting held 19–20 August, Apsara Angkor Hotel, Siem Reap, Cambodia.

- Mushi AK, et al. (2008). Development of a behaviour change communication strategy for a vaccination-linked malaria control tool in southern Tanzania. *Malaria Journal*: 7: 191 (doi: 10.1186/1475-2875-7-191.
- Panter-Brick C, et al. (2006). Culturally compelling strategies for behaviour change: a social ecology model and case study in malaria prevention. *Social Science and Medicine* 62: 2810–2825.
- Person B, D Cotton. (1996). A model of community mobilization for the prevention of HIV in women and infants. Prevention of HIV in Women and Infants Demonstration Projects. Public Health Rep; 111(Suppl 1): 89–98.).
- UNICEF-Regional Office for Southeast Asia (ROSA) (2006). *Behaviour Change Communication in Emergencies: A Toolkit*. Kathmandu: UNICEF-ROSA.

Finding out more

- Lee CI, et al. (2009). Internally displaced human resources for health: village health worker partnerships to scale up a malaria control programme in active conflict areas of eastern Burma. *Global Public Health*, 4:229–241.
- Mukanga D, et al. (2011). Can lay community health workers be trained to use diagnostics to distinguish and treat malaria and pneumonia in children? Lessons from rural Uganda. *Tropical Medicine and International Health* 16: 1234–1242.
- President's Malaria Initiative (2008). *PMI Communication and Social Mobilisation Guidelines*. http://pmi.gov/resources/publications/communication_social_mobilization_guidelines.pdf
- Richards AK, et al. (2009) Cross-border malaria control for internally displaced persons: observational results from a pilot programme in eastern Burma/Myanmar. *Tropical Medicine and International Health*, 14: 512–521.

CHAPTER 9

Operational research and associated routine monitoring

This chapter:

- describes how malaria operational research can help develop more effective responses in humanitarian emergencies
- identifies priority areas for operational research in humanitarian emergencies
- outlines planning and design of operational research

Importance of operational research in humanitarian emergencies

The essential interventions for malaria control in acute-phase emergencies are increasingly well defined, with malaria case identification and establishment of outbreak early warning systems as clear priorities. The choices of optimal preventive interventions, service delivery approaches, and behaviour change strategies are less certain and dependent on local context – which is why operational research is important.

Operational research, as defined here, is the search for information on strategies, interventions, or technologies that can improve the quality, effectiveness, or coverage of the control programme being studied (Zacharia et al. 2012). By addressing practical issues, operational research can guide intervention choices during specific emergencies, improving programme delivery and uptake by target populations. Additionally, good operational research can add to the evidence-base for malaria control programmes in humanitarian emergencies. There are many different approaches to malaria control, and no approach is 100% effective or appropriate in every situation. Thus, the purpose of operational research is to improve control tools, their application, and their impact.

Operational research should be distinguished from programmatic monitoring. Monitoring implementation of interventions (such as the extent to which patients with malaria are treated with effective antimalarials) and factors which may guide implementation of these interventions (such as the therapeutic efficacy of antimalarial treatments used) are routine activities for a well-managed malaria control programme. Operational research, in contrast, can address questions that arise in implementing an intervention in a specific context, such as identifying factors that influence treatment-seeking behaviour in a given emergency situation.

In the past decade, there has been increased recognition of the need for good-quality operational research in emergencies and of the role that research can play in improving malaria control in these settings. Operational research is no longer seen as the prerogative of external experts but is something in which many agencies are actively engaged, in identifying locally effective delivery of antimalarial treatments, appropriate prevention methods, and cultural and behavioural factors that affect control measures.

Because emergencies are fluid – moving from acute to early recovery or chronic situations – the feasibility and effectiveness of malaria control approaches will change as the situation progresses. Operational research can identify effective approaches in different phases of an emergency, and help guide necessary modifications as the situation changes. It can help in assessing the effectiveness of case management and prevention or outbreak control. Integration of operational research into normal control activities has helped solve malaria control issues arising in refugee settlements during the past decades (e.g. Afghanistan, Myanmar, Burundi).

The examples and issues identified in this chapter should encourage more agencies working in emergencies to conduct or participate in operational research and link with public health institutes and researchers who are engaged in improving health interventions in emergencies. An increasing number of humanitarian agencies now incorporate operational research into their relief efforts and have developed strong academic ties (e.g. Medecins Sans Frontiers, see http://fieldresearch.msf.org/msf/).

Guiding principles for operational research in humanitarian emergencies

Strict criteria must be met before operational research is conducted in humanitarian emergencies (see Table 9.1). As with all health research, the rights of individuals and communities must be protected and informed consent must be obtained. Approval from an independent and qualified ethics committee is necessary, which can be international if an appropriate local body is not available. Results must, where possible, be shared with stakeholders (e.g. study participants, government, NGOs, UN agencies) with the aim of improving local malaria control practices and informing necessary policy changes.

Acute and early recovery phases – when mortality reduction is the priority and resources may be stretched, operational research tends to be an opportunistic addition to operational priorities and should be:

- relevant and focused on local needs;
- as simple as possible;
- appropriately timed (e.g. as the situation may change before outcomes can be measured);
- designed to provide practical information to enable agencies to overcome operational problems and provide better interventions;
- planned and conducted in coordination with relevant government institutions (e.g. MOH) where feasible.

Chronic emergencies – where displacement camps and communities remain stable for several years, longer-term research is feasible and appropriate. It should be conducted as far as possible by local staff, with external support as necessary to build local skills.

Table 9.1 **Operational research considerations**

Relevance
How significant or severe is the problem? Can it be solved by other means that do not require research?
Avoiding duplication
Has the research already been done somewhere else? Was the context comparable?
Feasibility
How adequate are security, expertise, staff, time, and budget?
Support
Does the research have the interest and support of relevant stakeholders (e.g. MOH, donor, implementing partners)?
Applicability
Will the findings be applied and necessary resources made available?
Cost-effectiveness
Will the expected results justify the time, money and human resources invested? What difference will the study make to existing programmes?
Timeliness
Will the findings be available in time to inform decision-making?
Ethics
Is the research ethical? Will it benefit the individual participants and research subjects? Can informed consent be obtained?

Operational research areas

Two important areas of operational research for malaria control in humanitarian emergencies are improving the effectiveness of:

- diagnostic testing and treatment;
- prevention.

Other important areas include identifying implementation barriers and opportunities, social and behavioural constraints or opportunities, service delivery strategies, costs and cost-effectiveness of control activities.

Accurate diagnosis

Operational research can help assess the effectiveness of malaria diagnosis. Prompt and accurate diagnosis is critical but was often difficult in the acute phase of emergencies before RDTs became routinely available during the last decade. Clinical diagnosis of malaria is inaccurate and unreliable, while laboratory confirmation using microscopy is not always feasible in acute-phase or underfunded chronic emergencies.

RDTs are now considered an essential tool in non-emergency and emergency settings for diagnostic testing and estimating malaria prevalence. Despite the obvious advantages of malaria RDTs over clinical diagnosis, a number of operational challenges remain. Some useful operational research topics include:

- *Deploying RDTs in emergency settings* – If RDTs have not been fully scaled up in an area where an emergency is taking place, or if groups providing clinical care in the emergency setting have not previously used them on a large scale, they will need to be incorporated into clinical care protocols quickly. Previous experience in scaling up RDTs in non-emergency settings can be helpful, however, specific questions regarding scaling up a new diagnostic test in an emergency setting may need to be addressed through operational research.
- *RDT cost-effectiveness* – The scale up of RDTs is the priority in acute emergencies. In chronic emergencies, the role of microscopy and the cost-effectiveness of microscopy versus RDTs may need to be assessed, and there may be specific factors that influence the approach taken for the provision of diagnostic testing services. For information on collection of cost-effectiveness data see the ACT Consortium website (http://www.actconsortium.org/pages/guidance-notes.html).
- *Community usage of RDTs* – The optimal strategy for implementing RDTs among CHWs in an emergency setting may vary by setting, and operational research in this area may help guide programme decisions.

- *Clinical management of RDT negative febrile patients* – There is growing experience in the management of RDT-negative patients with febrile illness due to other causes, though there may be factors that are specific to their management in emergency settings. How these cases can best be diagnosed and treated will vary by location, and operational research can help develop guidance on this.

A range of methods can be used to conduct operational research into RDTs. Multidisciplinary approaches are often necessary, including clinical, parasitological, economic and social research. Social research is particularly useful in evaluating behavioural and attitudinal factors, such as how RDTs are being used in the field by clinic staff, or how patients and providers respond to negative RDT results.

Effective treatment

Plasmodium resistance to antimalarials as well as treatment provision issues (e.g. access to health facilities, treatment-seeking behaviour, adherence to treatment, drug quality, and quality of care) have an impact on the effectiveness of malaria treatment.

Drug resistance – During the last 20 years, resistance to sulfadoxine-pyrimethamine (SP) has become widespread in Africa and, more alarmingly, resistance to artemisinins has emerged in countries in the Greater Mekong subregion (Cambodia, Myanmar, Thailand, and Viet Nam as of 2012). The choice of antimalarial in the acute phase of an emergency should be guided by available information on efficacy of antimalarials in the area. Though typically considered part of programme monitoring rather than operational research, a therapeutic efficacy study may be considered in the acute phase of an emergency if clinical treatment failures raise suspicion of decreased efficacy of the chosen therapeutic agent. In chronic emergencies, two-yearly therapeutic efficacy testing in sentinel sites should be conducted as part of programme monitoring activities to ascertain that the chosen antimalarial agent remains effective (see chapter 4).

The *Global Plan for Artemisinin Resistance Containment* (GPARC) describes strategies to prevent, contain or eliminate artemisinin resistance through stopping its spread, increasing monitoring and surveillance, improving access to diagnostics and rational treatment, increasing action and resources, and investing in related research. GPARC research priorities include operational research to improve the field effectiveness of current and new tools, interventions and programmes for combating artemisinin resistance.

Certain emergency situations may be appropriate settings in which to explore some of the operational research priorities described in GPARC.

Treatment provision – Operational social research can help identify both the factors that prevent people from receiving effective malaria treatment and the changes that are needed in order to improve treatment quality. For example, it can help improve treatment access (e.g. by assessing whether health facilities are easily accessible and located in appropriate sites, or whether socially vulnerable groups have equal access to treatment) and treatment-seeking behaviour (e.g. whether people prefer private or public sector malaria treatment and why). This type of operational research can complement ongoing activities that monitor standards of care, to ensure that the quality of treatment provided by MOH, NGO and private health facilities is adequate and that health workers adhere to recommended drug treatment protocols. Surveys of public and private-sector coverage and prescription practices are recommended in every phase of an emergency. Such research, in conjunction with monitoring of quality care provision, and followed up by training and accreditation of competent practitioners, can help to improve standards, enable people to make informed choices about health providers, and assist the MOH to regulate standards in both public and private sectors.

Drug quality assurance – Poor quality drugs are a major problem that can result in patient death if unrecognized. Poor quality can be due to poor manufacturing practices or deliberate and sophisticated counterfeiting. Trademarks can be forged, so it is dangerous to buy drugs from nonofficial sources. ACT quality can only be assured by obtaining supplies from reputable, WHO-prequalified pharmaceutical companies, preferably directly from producers. Drug quality assessment would usually be considered a routine monitoring activity, most often carried out by those with specialized expertise in this area. In emergencies, where ensuring a supply of drugs may require purchase from local markets or other sources of uncertain quality, drug quality testing may be approached as an operational research study.

Standard protocols exist for collecting, documenting, storing and testing drugs from the formal, informal and private sectors. Advanced testing using High-Performance Liquid Chromatography (HPLC and mass spectrometry requires specialist laboratories. Some field-based laboratory kits (e.g. the GPHF-Minilab®) can detect seriously flawed drugs that contain either the wrong ingredients or no active ingredients.

Effective prevention

Operational research can help assess the effectiveness of prevention measures in a particular setting, playing an important role in evaluating transmission control and personal protection interventions during emergencies. It can help identify ways to improve delivery and usage (e.g. social research on community preferences, and testing coverage achieved by different delivery methods). With the rapid scale-up of LLIN coverage globally, resistance to pyrethroid insecticides is increasing, and insecticide resistance testing may be necessary in many emergency settings. More detail on entomological monitoring and insecticide susceptibility testing is provided in Chapter 7.

LLINs – Research has shown that LLINs provide personal protection in many settings and can provide additional benefits in high-endemicity settings such as in Africa, if population coverage is high, through a mass killing effect on mosquito populations (see Chapter 7). There is no longer justification to use mosquito nets other than LLINs. There are few endemic areas where ITNs or LLINs have not been tested, and the areas where they work well are mostly known.

LLIN durability during emergencies is a current operational research issue. LLINs are expected to last for at least 2–3 years in non-emergency settings. Recent data from a number of endemic countries in Africa, suggest that the lifespan of certain LLINs may be less than 2 years in some settings. In emergency settings, their lifespan is likely to be shorter, but how much shorter has not been documented. LLINs are currently made from polyester, polyethylene or polypropylene fibres; it is not known which is more suited to emergency conditions. WHO has developed standardized guidelines on assessing the durability of LLINs (residual efficacy and physical integrity), and information from emergency settings would be a valuable addition (WHO, 2011).

LLINs work well against all African malaria vector species, *if used regularly and correctly*, but are not effective in every epidemiological context. For example, many South-east Asian vector species bite predominantly in the evening and morning (e.g. *An. minimus*) and LLINs can only protect against malaria while people are sleeping under them. Additional research on malaria impact and adjunct protective strategies in such settings is needed. LLINs may be less effective in acute emergencies, unless communities are accustomed to using mosquito nets and shelters are suited to hanging them. However, evidence exists of their effectiveness in post-acute phase and in chronic emergencies (Kolaczinski et al., 2004). Ideally, before LLIN coverage is expanded in an emergency setting, a rapid assessment should be conducted to:

- determine previous experiences with using mosquito nets within the displaced population, whether people want or would use LLINs, and the importance of protection against mosquito bites relative to other priorities;
- determine appropriate colour, sizes and shapes, a frequently neglected component of choosing the right LLIN brand;
- develop appropriate health communications materials on net usage and care, including repairs by users;

As no alternatives to pyrethroids yet exist for LLINs, their disease-control impact should be monitored regularly, particularly in areas with pyrethroid or organochlorine resistance. Local resistance to pyrethroids can be monitored using WHO susceptibility test kits or CDC bottle bioassays. Procedures are described in the WHO revised test procedures for insecticide resistance monitoring (WHO, 2013). Operational research into these and other basic questions can help in determining the most cost-effective ways of achieving and maintaining high LLIN coverage and usage in post-acute or chronic emergency settings.

IRS – This has become more common for community malaria control during emergencies in certain countries. In areas of pyrethroid resistance, and where LLINs are also being deployed at high coverage, WHO recommends that IRS with pyrethroids be replaced with alternative insecticides (e.g. carbamates, organophosphates), to curtail further selection of pyrethroid resistance. IRS with any insecticide requires coverage of approximately 80% of dwellings to be effective and is generally not feasible during acute-phase emergencies (see Chapter 7). Prior to insecticide procurement, susceptibility of local vector populations should be tested using either WHO tube assay (see Annex X) or CDC bottle assay (http://www.cdc.gov/ncidod/wbt/resistance/assay/bottle/index.htm). After insecticide is applied, it is desirable to monitor the duration of effectiveness using a WHO cone bioassay on live susceptible mosquitoes (Annex X) or one of the colorimetric assays under development.

Operational research has helped to formulate guiding principles on IRS, but more research is needed to determine how effective IRS with carbamates or organophosphates is and whether IRS with these compounds can have a sustained impact on malaria transmission in emergency settings.

Alternative protection methods

The utility of LLINs in emergencies should not be assumed if their acceptability is uncertain, and they may be less practical when populations are living in tents or other types of emergency shelter. In this case, operational research may be needed to assess the potential of alternative approaches, such as pyrethroid treatment of blankets/topsheets and ITPS (see Chapter 7). One potential advantage is that these materials are already being distributed in acute emergencies, so if they could be rendered protective against malaria (in areas where local vectors are still susceptible to pyrethroids) this would provide dual benefits without placing extra demand on logistics. One study in West Africa indicated that, as with IRS, ITPS may provide community protection through a mass killing effect on mosquito populations (Burns et al. 2012). There is also good evidence from Asian and African trials for the effectiveness against malaria of *in situ* insecticide treatment of blankets/topsheets with permethrin in displacement settlements (Rowland et al., 1999).

While there is some evidence to suggest the use of such materials in emergencies, the evidence base is still weak compared to that for IRS and LLINs – which includes many community-randomized trials in different settings against different vectors. Commercial permethrin-treated blankets have not yet been tested in trial settings. Additionally, there have been no trials to date of ITPS and insecticide-treated blankets used together, despite both blankets and plastic tarpaulins being routinely distributed in emergencies for warmth and shelter. Both ITPS and permethrin-treated blankets are now produced commercially. However, at present there are not sufficient data for WHO to make a recommendation regarding these interventions. However, they are included in this handbook because they are currently being used in the context of humanitarian emergencies. Further research is required to assess:

- the effectiveness of treated topsheets and blankets in different cultural, epidemiological and climatic settings in Africa and Asia;
- the potential of ITPS in settings other than West Africa;
- whether there is increased effectiveness when the ITPS and permethrin-treated blankets are used together as compared to either intervention being used alone.

IPT – More than a decade ago, intermittent preventive treatment emerged as a new strategy, initially for use among pregnant women (IPTp), and then more recently for use in infants (IPTi). IPTp is normally delivered through antenatal care. Deployment of IPTp through generally weak African health systems has proven challenging and is the focus of operational research. As routine antenatal care may be weak or non-existent during humanitarian emergencies, research may be needed on alternative delivery strategies. Information on the effectiveness of IPT in these settings – measured as prevention of severe anaemia and low birth-weight – in areas where it is implemented, could inform the broader debate about use of IPT in areas with relatively high levels of SP resistance (see Chapter 6).

Seasonal malaria chemoprevention (SMC) is now recommended for incorporation into malaria programmes in countries with seasonal malaria and low resistance to SP. A complete treatment course of amodiaquine plus SP should be given to children aged between 3 and 59 months at monthly intervals, beginning at the start of the transmission season, to a maximum of four doses during the malaria transmission season. This recommendation is currently limited to the Sahel.

No major studies of IPT or SMC have been conducted in emergency settings. However, both are logistically feasible in post-acute phase emergencies as treatments are intermittent. Seasonal chemoprevention could be expected to be useful for other target groups (e.g. displaced populations), but has not yet been evaluated for this purpose.

Effective and sustainable delivery

While operational research is important to identify the best delivery strategy during the different phases of emergencies, this should not be an excuse for not providing free services to attain universal coverage of interventions. It is therefore generally accepted that malaria services must be free-of-charge to end-users during emergencies. Operational research can help to address questions related to maintaining effective delivery during changing emergency phases and conditions. These questions could include:

Accessibility

- Which delivery strategies achieve best coverage and uptake during acute-phase emergencies?

Demand creation

- What is the best way to encourage appropriate use of prevention and treatment services?
- Which communication methods work best in different emergency phases (e.g. mass media, person-to-person)?

Monitoring

- How can behaviour change and correct usage best be monitored?
- What are the most effective behaviour change communication strategies for different emergency phases?
- Does the inclusion of community focal persons (e.g. religious leaders, teachers) improve malaria treatment-seeking and use of preventive measures?
- Does investment in lay health-workers (e.g. CHWs) in acute-phase emergencies improve malaria outcomes?
- What are the impacts of involving CHWs in case management?

Effective outbreak control

Surveillance can also be used as an operational research tool to evaluate the impact of outbreak control interventions (see Chapters 4 and 5). For example, it was suspected that IRS with lambda-cyhalothrin was no longer working effectively in Afghan refugee camps, but cross-sectional prevalence surveys before and after spraying confirmed that it was still controlling malaria effectively (Rowland et al., 1994). Other indicators that might be used for measuring control impact, depending on the resources available, include the number of malaria cases presenting at clinics, prevalence of anaemia, and mosquito population density.

Social and behavioural determinants of intervention uptake

Successful malaria control depends on community members understanding and using prevention and treatment measures. It is vital to use information on community beliefs and practices about malaria to improve uptake of interventions. Social and behavioural research is important for designing effective interventions and can help explain why a particular intervention fails to be accepted.

Operational research on social and behavioural factors influencing intervention uptake can strengthen planning, implementation, and monitoring by providing information about people's beliefs and prevention and treatment-seeking behaviours (see Chapter 7 and Annex IX). Questions can include:

Perceived importance of malaria and prevention

- What perceptions do people have about causes of febrile illness?
- What perceptions do people have about causes of malaria?
- How do people traditionally protect themselves from mosquitoes?
- How do people traditionally protect themselves from malaria?

- What measures are currently acceptable (e.g. LLINs, IRS, ITPS)?
- What household items do people most value (e.g. are LLINs valued or are they sold)?
- Who makes decisions about family protection/health care and could be targeted by health communications?
- What are family sleeping arrangements (e.g. co-sleeping, shared rooms, average household size)?
- If LLINs are used, who sleeps under them (e.g. gender or age differences in LLIN usage)?
- If LLINs are used, how are they washed and how often?
- Does help with hanging of LLINs improve usage rates?

Treatment-seeking and access

- What influences treatment-seeking (e.g. who in the family decides, what is the decision process)?
- What local or traditional treatments are used?
- Where is treatment obtained (e.g. public services, private/informal practitioners)?
- When is treatment sought (e.g. is there delay, for how long, are traditional treatments tried first)?
- What access do people have to effective antimalarials (e.g. does the private sector provide quality-assured ACTs, how is the supply of drugs regulated or maintained, do traditional/informal practitioners prescribe antimalarials)?
- Do most people accept recommended malaria treatment? Is local antimalarial treatment policy different from their former experience of treatment?
- How do people feel about antimalarial safety and efficacy?
- What are barriers to treatment and care (e.g. are services accessible, what role do intermediaries – such as other family members – play in accessing care)?
- Do security concerns affect treatment-seeking or access to care?
- Are some social groups excluded because of physical or political factors (e.g. gender, age, and ethnicity)?

Social and behavioural information can be collected through standardized social research methods, including key informant interviews (e.g. with community leaders, CHWs), in-depth interviews with community members, household surveys, and participatory action research (see Chapter 7). Each of these methods is feasible in emergency settings, though large-scale community surveys in particular pose logistical challenges and require technical assistance.

Economic research

The costs and cost-effectiveness of malaria control in acute-phase emergencies are still poorly documented. Economic costing analyses of malaria control in emergencies are feasible, particularly using a provider perspective, because most agencies maintain detailed expenditure records for financial reporting to donors. More research is also needed to identify prevention and treatment approaches that will be financially sustainable post-emergency, after the withdrawal of humanitarian aid.

Operational research planning

Planning operational research

Malaria operational research and malaria control require similar processes of planning, design, choice of tools and indicators, as well as monitoring and evaluation. However, operational research differs from malaria control in that it starts from a position of uncertainty and aims to answer a specific question or identify the best solution to a problem – which may involve comparing results from a number of approaches. The steps needed in operational research remain the same whether the aim is to improve an existing malaria control strategy or to choose among available interventions (see Table 9.2).

Good operational research planning and implementation are vital. If activities are not conducted properly or an outcome is not recorded properly, results may be incorrect and the interpretation of findings will be difficult or impossible. Important aspects of planning include:

- agreeing on clearly-defined objectives and outcomes;
- planning time-bound activities (e.g. including training data collectors and piloting);
- establishing a systematic approach to evaluating outcomes with appropriate technical and process indicators;
- designing for evaluation from the beginning of study development.

It is necessary to plan at the design stage how data will be collected and analysed. Planning data collection helps to clarify activities (e.g. data collection, quality control checks, data management, analysis), minimize errors, avoid unnecessary delays, and organize human and material resources. Planning how data will be analysed helps to formulate clear and specific objectives and determine what information should be collected in the first

Table 9.2 **Steps in planning operational research**

Questions	Next steps	Actions
What is the problem?	Formulate the research question.	Use existing literature to strengthen development of the research question.
What information is already available?	Review published and grey literature.	Review clinical and/or social science literature, government reports, agency records.
What is the purpose?	Formulate objectives.	Develop 3–5 SMART objectives that can answer the research question.
What additional data are needed and how should data be collected?	Develop research methods.	Study type, study design, population, sampling, ethics, piloting, data collection, analysis, dissemination.
How will results be used?	Determine how findings can inform implementation.	Do not collect research data without knowing how the results will be used to improve implementation. It is helpful to involve potential beneficiaries in research planning and implementation.
Who will do what and when?	Develop work plan.	Appoint and train research staff, schedule implementation, monitoring and supervision.
What resources are needed?	Develop budget and resource allocation plan.	Mobilize funding, recruit staff, and procure necessary supplies and equipment.
How should research question and findings be disseminated to donors, authorities, partners, and communities?	Develop dissemination plan.	Establish a committee of experts to review the implementation of the dissemination plan including the sharing of research results.

place. Pre-testing a particular research component or piloting the full methodology with a small sample helps identify potential problems, allowing necessary revisions before research is implemented on a larger scale.

Designing operational research

Research conducted in humanitarian emergencies includes:

- *Research that is specific to local issues* – which may need to be investigated in every emergency (e.g. cultural beliefs about disease causation, health-seeking behaviour);

The distinction between operational research in humanitarian and non-humanitarian settings is not always clear and humanitarian agencies may be involved in both. For example, after a new prevention or treatment intervention is developed it may need to be evaluated in emergency settings with different epidemiological and social conditions. The main approaches used in operational research are qualitative, quantitative, or a mixture of the two.

Qualitative research usually focuses on small purposively selected samples, to allow variables to be explored in depth. It is particularly useful for exploring attitudes, behaviours, and motivations, or characteristics of particular situations that may change over time. For example, qualitative research among Afghan families explored health attitudes and approaches to malaria prevention within the social constraints provided by the Taliban regime (Howard et al. 2010). It is often used as a precursor to quantitative research as it can help in developing a testable research question.

Quantitative research is the standard experimental scientific approach, often using population samples and statistical analysis. The quantitative study of incidence, distribution and control of diseases such as malaria is called epidemiology. It can be used for disproving discrete research hypotheses. For example, a quantitative study in Afghanistan showed the association between increased insecticide-treated net coverage and reduced malaria morbidity (Rowland et al., 2002; see Box 9.1).

Quantitative operational research uses epidemiological principles and can be either descriptive or analytical. *Descriptive studies* use cross-sectional surveys or longitudinal follow-up of a cohort to systematically observe and describe the group or situation without attempting to influence outcomes (e.g. the frequency of malaria cases in a population by age or population group). Sources of descriptive data in emergencies include surveillance, health facility records, and household surveys. *Analytical studies* aim to investigate causal relationships between malaria mortality or related outcomes and various exposures. These can be either observational or intervention study designs. Observational study designs can be ecological, cross-sectional, cohort, or case-control. Cross-sectional (e.g. for descriptive analysis), case–control, and retrospective or prospective cohort studies are most commonly used in operational research. Intervention studies can be considered for particular operational research questions if the skills and resources are available to conduct them well.

- *Cross-sectional studies* relate exposure status and outcome prevalence at one time-point from a random sample of the population of interest. For example, prevalence of severe malaria associated with living in a particular displacement camp. They are relatively quick and easy, though it is difficult to know if exposure preceded outcome, and are often used in operational research to provide preliminary evidence of an association.

- *Cohort studies* compare individuals with and without exposure to a risk factor to measure the occurrence of the outcome of interest over time. For example, measuring the incidence of severe malaria in pregnant and non-pregnant women. Cohort studies in which subjects are followed longitudinally can be time-consuming, labour-intensive and expensive, making them more appropriate for stable settings (e.g. long-term displacement camps). Retrospective cohort studies, in which exposures and outcomes are assessed after they have occurred, are relatively less resource intensive and may be employed in some acute settings.
- *Case–control studies* select individuals on the basis of outcome status and analyse whether they differ in terms of previous exposure. For example, malaria mortality among those who used ITPS versus those who used untreated tarpaulins. Case–control studies are relatively quick and inexpensive, and are therefore often favoured for outbreak investigations and estimating the impact of an intervention in emergency conditions.
- *Intervention studies* allocate a protective factor to individuals or groups and compare the frequency of the outcome of interest between those exposed and those unexposed. For example, comparing incidence of treatment failure between those given a new treatment drug and those given current treatment. Intervention studies can be slow, labour intensive and expensive, but are the gold standard – particularly double-blind randomized controlled trials – for inferring causality with protective exposures.
- *Mixed-methods research* combines elements of qualitative and quantitative approaches. It is often used for addressing policy-focused questions.

Selecting exposure and outcome measures

The main outcomes of interest in malaria operational research are:

- disease incidence
- disease prevalence
- intermediate variables (e.g. vector density, human behaviour change).

In studies of malaria control interventions in emergencies, the difference in incidence or prevalence of disease in the intervention and control groups is the most relevant outcome. Once a disease surveillance and monitoring system is in place, it is simpler and more definitive to measure changes in disease outcome than changes in intermediate outcomes (e.g. changes in knowledge or behaviour). However, intermediate outcomes can be useful for refining or improving an intervention or explaining how it works.

Exposures of interest can depend on study purpose and design. The main exposure of interest could be usage of a protective item (e.g. LLINs, IPT, ACT). Where intermediate variables are particularly technical, specialist advice should be sought from international or local experts.

Sampling

Quantitative operational research most often uses a sample population (e.g. of individuals, households, camps) and extrapolates findings to the target population (e.g. the crisis-affected population). To reduce selection bias, the study population should be representative, possessing all characteristics of the target population from which it is drawn. To reduce the role of chance, the study population should be sufficiently large (e.g. if the study is too small it may estimate effects too inaccurately and if it is too large it will use more resources than necessary).

Two important statistical concepts in calculating sample size are power and precision. Power is the probability of detecting an effect if it is real, while precision is the probability of detecting an effect if it is not real. A study should typically have 80–90% power and 5% precision (e.g. shown by p-values) to accurately detect an effect. Computer programmes can be used to calculate required sample sizes (e.g. EpiInfo, SPSS). In emergency settings, the sample size may need to be increased to allow for relatively high losses to follow-up. Losses to follow-up are likely in displaced and crisis-affected populations. For example, during a 1-year study of insecticide-treated topsheets in Kabul, Afghanistan, more than 30% of the study population moved because of security problems. The higher the rate of loss to follow-up, the less robust study results will be. Other factors that determine the size of a study include availability of staff, transport, laboratory capacity, time and money.

Once sample size has been calculated, subjects must be selected (i.e. sampled) from the target population. Subjects might be individuals (e.g. in drug trials), households (e.g. in an LLIN trial) or communities (e.g. in an IRS trial). Subjects are allocated randomly to eliminate potential bias due to unforeseen factors. The double-blind approach, in which neither the researcher not the participant is aware of who is allocated to the intervention or control group, eliminates potential bias in the assessment of intervention impact, but is not possible for all types of interventions. Random sampling requires a sampling frame, or list, detailing all eligible members of the target population for selection. Sampling methods include:

- *Simple random sampling*, in which each subject in the sampling frame has an equal chance of being selected (e.g. through random number tables). A simple random sample may be difficult because of logistics or problems of excluding neighbours;
- *Systematic sampling*, in which subjects are selected at regular intervals (e.g. every fifth household) can avoid some of the difficulties with simple random sampling. Stratified sampling selects representative subgroups (e.g. by age, ethnic groups) in proportion to the population structure.

If a sampling frame is not feasible or the population is widespread, other methods such as stratified, multi-stage, or cluster sampling may be used:

- *Cluster sampling*, in which a random sample of villages or camps is chosen and all target individuals (e.g. children under age five) are selected for

Box 9.1 **Case–control study in Afghanistan**

The impact and coverage of insecticide-treated nets can be estimated using routine data collected by passive case detection at field clinics. Data from the microscopy registers of Behsud clinic in eastern Afghanistan were as follows:

ITN user status	Falciparum-positive (cases)	Falciparum-negative (controls)	Total
ITN user Non-user	52 (a) 439 (c)	610 (b) 1958 (d)	662 (e) 2397 (f)
ITN non-user	439 (c)	1958 (d)	2397 (f)
Total			3059 (g)

These data were used to estimate individual effectiveness through a case–control study. Those with slides positive for *P. falciparum* were considered to be cases, and those with slides that were negative were considered to be controls. The odds ratio (ad/bc) = 0.38, and individual protective effectiveness is 1–odds ratio = 0.62 or 62% (confidence interval 48–72%). Data were also used to estimate **the proportion of the population using ITNs** (e/g) = 0.22.

Thus, the impact on malaria morbidity in the population, or **community effectiveness** = individual effectiveness multiplied by coverage, which in this community is 0.62 x 0.22 = 14%. It was concluded, therefore, that 14% (confidence interval 11–16%) of the total number of malaria cases that would have occurred in the community, were prevented by the ITN distribution programme (Rowland et al., 2002).

the intervention, is logistically feasible but more likely to be unrepresentative. This risk can be reduced if the number of clusters is 20 or more.

Humanitarian research ethics

Operational research incurs ethical obligations, particularly in humanitarian settings when working with vulnerable displaced and crisis-affected communities. Most operational research for malaria control is either etiological (e.g. what caused the malaria outbreak?), systems-related (e.g. which LLIN delivery strategy works better, which insecticide is most cost-effective?), or clinical (e.g. is this intervention effective?), and is addressed following the principles of biomedical and social research ethics.

Guiding principles – WHO policy on research ethics states that all research involving human participants must be conducted in a manner that respects the dignity, safety and rights of research participants and recognizes the responsibilities of researchers. The fundamental bioethical principles to apply are autonomy (i.e. right of refusal), beneficence (i.e. in-line with participants' best interests), justice (i.e. equitable distribution of benefits), non-maleficence (i.e. do no harm), and human dignity (i.e. respect for persons).

All operational research should follow the guiding principles in the amended *Helsinki Declaration* of 1989, further elaborated for developing countries in the updated *International ethical guidelines for biomedical research involving human subjects* (CIOMS, 2002) and *International Ethical Guidelines for Epidemiological Studies* (CIOMS, 2008). The Economic and Social Research Council research ethics framework is particularly useful for social research (ESRC 2012).

Ethical considerations – People affected by emergencies are particularly vulnerable. Key considerations before implementing operational research include:

- purpose (e.g. who will benefit most – the community being studied, the global scientific community, or the researcher?);
- selection (e.g. particular topics and participant groups, such as children under five and women, are studied more frequently than others and sometimes in accordance with research rather than community priorities);
- power (e.g. power imbalances such as differing social status of researchers and participants can affect consent, participation, validity of responses, and even outcomes);

- risks (e.g. unintended consequences can be both social and psychological, including recrimination, security breaches, stigmatization, and raised expectations);
- consent (e.g. can participants understand study purpose, beneficiaries, implications of involvement, and choose whether or not to participate?);
- confidentiality (e.g. concepts of privacy and participant identifiers can vary considerably);
- dissemination (e.g. results that are not both locally and broadly disseminated can lead to duplicated efforts, sometimes by different researchers using the same participants).

Operational research that provides a generalized benefit to populations affected by humanitarian emergencies is essential, but should not be conducted without benefit to the community being studied. Similarly, vulnerable socially and economically deprived individuals must not be used in research that will mainly benefit more privileged individuals. Intervention trials on refugees should be undertaken only when there is uncertainty about potential benefit. If a previous study provided evidence for a benefit, there is no justification for a trial unless there is good reason to believe that the results might be different in the present population or locality.

Community participation is essential for most operational research. It is important to involve community leaders and representatives in research planning and design. Informed consent must be obtained from all participants and it is the investigator's responsibility to ensure that participants are fully informed of the potential risks and benefits of participating in a study. Difficulties in ensuring informed consent include individual autonomy, hospitality norms, power imbalances and dependency. For example, individual participants may feel pressured by cultural expectations of politeness, by the hope of some tangible community benefit, or by community leaders' expectations of their cooperation.

Choice of control in intervention studies is difficult. The norm is that any new intervention should be compared with the best intervention currently available. Comparison with a placebo or a "no intervention" control group is acceptable only if no other effective intervention is known. Thus, if an intervention is already being used, its withdrawal for research purposes is unethical. However, if it is not clear whether an intervention in use remains effective, a randomized trial comparing a promising new treatment with the existing treatment is justified. If, as is often the case, there are insufficient resources to meet the needs of everyone, distribution might be randomized to enable an unequivocal assessment to be made. Additionally, it is acceptable to try a new intervention that is generally equivalent to the existing intervention if it is cheaper, easier to implement, more sustainable, associated with fewer adverse reactions, or more acceptable to the community. It may be preferable to conduct comparative trials rather than randomized controlled trials to avoid the risk of withholding a beneficial treatment.

Ethical review procedures

Any research protocol that involves people should pass formal review by an ethics review committee. This is not only a basic prerequisite for journal publication, which is important for dissemination, but helps ensure that ethical ramifications have been addressed. Ethics review can be particularly challenging in emergency settings as local ethics review boards may not exist and those traditional boards that do exist (e.g. in academic institutions) may not be able to provide timely approval. Some humanitarian organizations (e.g. MSF) have set up their own independent ethics review boards to help address this.

Data collection and analysis

Data collection and analysis are not covered in this handbook. Since both of these often require specialist support, it is important to seek advice before data collection begins to ensure that the study adds to the evidence base rather than wasting time and resources.

Disseminating and using findings

The ultimate relevance of operational research is whether it contributes to improving the effectiveness of interventions or influences policy change (Zacharia et al 2012). Dissemination requires identifying and targeting messages to an appropriate audience. The audience should include the community in which operational research was conducted and local and/or international policy-makers depending on the generalizability of the research findings. Whether research findings are positive or negative, they must be reported to national health officials in the MOH, donor agencies, community leaders and participants, and implications for policy discussed. Increasingly, operational research is disseminated more broadly through publications in peer-reviewed scientific journals.

Using research findings to guide programme development is critical. Successful new interventions may need to be integrated into current control programmes, and strategies to do this may need to be developed.

References

- Burns et al. (2012). Insecticide-treated plastic sheeting for emergency malaria prevention and shelter among displaced populations: an observational cohort study in a refugee setting in Sierra Leone. *Am J Trop Med Hyg*. 2012 Aug; 87(2): 242–50. doi: 10.4269/ajtmh.2012.11-0744.
- CIOMS (2002). *International ethical guidelines for biomedical research involving human subjects*. Geneva, Council for International Organizations of Medical Sciences.
- CIOMS/WHO (2008). *International Ethical Guidelines for Epidemiological Studies:* Council for International Organizations of Medical Sciences (CIOMS) and the World Health Organization (WHO): Geneva, Switzerland.
- ESRC (2012). *Framework for Research Ethics (FRE) 2010*, updated September 2012. (http://www.esrc.ac.uk/_images/Framework-for-Research-Ethics_tcm8-4586.pdf)
- Howard N, Shafi A, Jones C, Rowland M (2010). Malaria control under the Taliban regime: insecticide-treated net purchasing, coverage, and usage among men and women in eastern Afghanistan. *Malaria Journal*, 9(1), 7.
- Kolaczinski JH et al. (2004). Subsidized sales of insecticide-treated nets in Afghan refugee camps demonstrate the feasibility of a transition from humanitarian aid towards sustainability. *Malaria Journal*, 3, 15.
- Rowland M, Hewitt S, Durrani N (1994). Prevalence of malaria in Afghan refugee villages in Pakistan sprayed with lambdacyhalothrin or malathion. *Transactions of the Royal Society of Tropical Medicine and Hygiene*, 88(4): 378–379.
- Rowland M et al. (2002). Prevention of malaria in Afghanistan through social marketing of insecticide-treated nets: evaluation of coverage and effectiveness by cross-sectional surveys and passive surveillance. *Tropical Medicine and International Health*, 7:813–822.
- Rowland et al. (1999). Permethrin-treated chaddars and top-sheets: appropriate technology for protection against malaria in Afghanistan and other complex emergencies: *Transactions of the Royal Society of Tropical Medicine and Hygiene*, volume 93, Issue 5, Pages 465–472, September 1999.
- WHO (2011). *Guidelines for monitoring the durability of long-lasting insecticidal mosquito nets under operational conditions*. World Health Organization.
- http://whqlibdoc.who.int/publications/2011/9789241501705_eng.pdf.

- WHO (2013). *Test procedures for monitoring insecticide resistance in malaria mosquitoes.* Geneva, World Health Organization. http://www.who.int/malaria/publications/atoz/9789241505154/en/index.html
- Zachariah, R., et al. (2012). Is operational research delivering the goods? The journey to success in low-income countries. *Lancet Infect Dis*, 12(5), 415–421. doi: 10.1016/S1473-3099(11)70309-7

Finding out more

- Carneiro I and Howard N (2011). *Introduction to Epidemiology*. Open University Press.
- Chandler, CIR (2009). ACT *Consortium Social Science Guidance*. ACTC/CC/2009/SSGv04.
- Dondorp AM et al., (2009). Artemisinin resistance in Plasmodium falciparum malaria. *N Engl J Med*. 361(5): 455–67.
- FIND RDT quality control: http://www.finddiagnostics.org/programs/malaria/find_activities/rdt_quality_control/
- Lubell Y et al. (2008). An interactive model for the assessment of the economic costs and benefits of different rapid diagnostic tests for malaria. *Malar J*. 2008 Jan 28;7:21.
- Mills, E. J, & Singh, S (2007). Health, human rights, and the conduct of clinical research within oppressed populations. *Global Health*, 3, 10. doi: 10.1186/1744-8603-3-10
- Newton PN et al. (2009) Guidelines for field surveys of the quality of medicines: A proposal. *PLoS Med* 6(3): e1000052. doi:10.1371/journal.pmed.1000052.
- Phyo AP et al. (2012). Emergence of artemisinin-resistant malaria on the western border of Thailand: a longitudinal study. *Lancet*, 379:1960–1966.
- WHO (2011). *Global plan for artemisinin resistance containment (GPARC)*. Geneva, World Health Organization. http://www.who.int/malaria/publications/atoz/9789241500838/en/index.html
- WHO (2009). *Methods for surveillance of antimalarial drug efficacy*. Geneva, World Health Organization. http://www.who.int/malaria/publications/atoz/9789241597531/en/index.html
- WHO (2003). *Operational research for malaria control. Learner's guide and Tutor's guide* (trial edition). Geneva, World Health Organization, 2003 (WHO/HTM/RBM/2003.51 Part I and II).

Annexes

ANNEX I

Sources and methods for collecting population data

Counting households

The total size of a population is estimated by multiplying the number of households by the average number of people per household: this information can be obtained by exhaustive counting or systematic sampling.

Exhaustive counting of households

1. Count the total number of households in the area. This can be done on foot, from a vehicle, or by aerial photography. Because it is exhaustive, this method is most appropriate for small sites covering limited areas.
2. Calculate the average number of persons per household by conducting a small survey of sample households selected at random. A minimum of 30 households should be selected.
3. Estimate the total population by multiplying the total number of households by the average number of persons per household.

Systematic sampling

1. This technique is particularly adapted to well-organized refugee camps. Using interval sampling and a departure number chosen at random, select a sample (and thus determine each household to be visited). This method assumes that households are arranged in such a way that interval sampling is possible and that their approximate number is known.

   For example:

   Estimated no. of households = 4000

   Convenient sample size chosen = 400

   Sample interval = 4000/400 = 10 (information on the number of person living in an household will be collected in every 10 habitats)

   If the randomly chosen departure point is 6 (i.e. the sixth household beginning at one extremity of the camp), the selected households are therefore number 6, then number 16 (6+10), then number 26 (16+10), etc.

2. Estimate the total population by multiplying:
   (total no. of households visited) x (average no. of persons per household) x sample interval.

Mapping

The total size of a population is estimated by multiplying the total area of a site (m^2) by the average population density per m^2: this information can be obtained by the quadrate method or the T-square method.

Quadrate method

1. Draw the camp boundary. This can be accomplished either by taking GPS points along the perimeter and drawing the map with software, or by hand with a compass and then on paper.
2. Select 30 systematically random locations within the site. This can also be accomplished by mapping software or by hand. In Figure I.1, six such points are shown.
3. Mark off a 25 m x 25 m quadrate, or block, physically with a rope or by using a telemeter at each point.
4. Count the population within each quadrate, and the number of persons and number of households for the 30 quadrates.

Figure I.1 **Quadrate method for calculating population size**

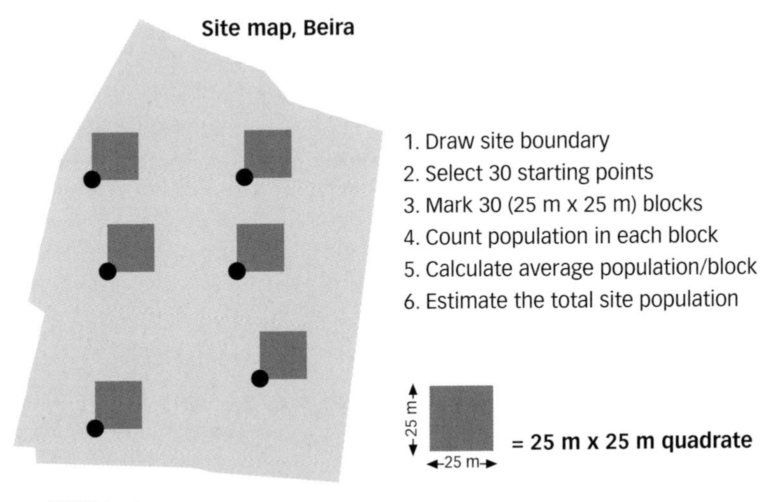

Source: MSF/Epicentre.

5. Calculate the average population per quadrate.
6. Estimate the site population by extrapolating the average population per quadrate to the entire site surface. Confidence intervals are then calculated around the estimate.

T-square method

The T-square method "involves sampling a number of random points, measuring the distance between each point and the nearest household or family unit, and then measuring the distance between that household and the next closest one, as a way of measuring population density". It gives a more accurate evaluation of the population size, but is more complicated to perform than the quadrate method.

Census or registration

Census

A census of the displaced population is the ideal method for ascertaining population data, if it is feasible. It involves visiting homes and counting how many people live in each. If security allows, it is best done early in the morning or in the evening when refugees or displaced people are more likely to be "at home". However, as the following example illustrates, in an emergency situation, there may not be sufficient time or human resources to carry out a census.

> Following the floods in Mozambique, a quick census was carried out by an agency in two temporary accommodation centres. The census was done at night when people were most likely to be at home. It was found that, although the number of residents had been estimated at 10 000, the actual number of people residing in the accommodation centres was around 6000. The reason for this discrepancy was that many people who were present in the camp during the day were actually returning to their villages at night.

Registration

Registration of refugees or displaced people when they arrive at a site can provide an opportunity to collect data about population size (and structure). Registration can also be combined with other activities, such as distribution of food cards, detection of malnutrition and vaccination against measles.

Data from programme activities

Information from programme activities, such as a vaccine coverage survey of a specific age group (e.g. children aged 6–59 months), can be used to estimate the number of children in this age group as well as the total population.

For example:
A measles vaccination survey estimates vaccination coverage among children aged 6–59 months in a camp to be 80% (0.8) and 5000 measles vaccines were administered in this age group. Using these data, the number of children aged 6–59 months can be estimated: 5000/0.8 = 6250. If, from another survey, it is known that children in this age group represent 20% of the population in question, the total population can be estimated: 6250/0.2 = 31 250.

Population structure

Since vulnerable groups need to be monitored, it is useful to know what proportion of the population are pregnant women and children under five. Population structure can be estimated based on the typical distribution of age groups in the general population.

Table I.1 **Distribution standard by age of stable populations in developing countries**

Age group	Proportion of total population
0–4 years	17%
5–14 years	28%
15–29 years	28%
≥30 years	27%
Total	**100%**

Humanitarian emergencies, however, can affect the normal age and sex structure of populations. For example, there may be proportionally fewer men and proportionally more women, young children and elderly people. It is therefore necessary to conduct a simple census or sample survey to find out about the age breakdown of the population.

As a minimum it is important to estimate the *expected* number of children under 5 years of age and pregnant women *if the camp population were of the same composition as a normal population*. Therefore, without a census of the population, it is reasonable to assume children under 5 years of age represent approximately 17% of the population. A quick way to estimate the number of pregnant women is to use the following calculation, with values based on typical values:

total number of pregnant women =

total population x

proportion of women in a population (typically 51%) x

proportion of women of childbearing age (15–45 years) among all women in the population (typically 50%) x

chance of any woman aged 15–45 years being pregnant for a given fertility rate[1] (approximately 20% for a fertility rate of 8)

Therefore, total number of pregnant women = total population x (0.51 x 0.50 x 0.20)

[1] In a given population, the total fertility rate (per women 15–45 years) may be, for example 5. If so, a woman is pregnant 5/30 years (16.7%) for 9/12 months (75%); from this, the chance of any woman aged 15–45 years being pregnant at a given time = 16.7% x 75% = 12.5%. At a total fertility rate of 8 children per woman of 15–45 years, 20% of women of childbearing age can be expected to be pregnant at any given time.

ANNEX II

Methods for collecting retrospective mortality data

Retrospective mortality data can help to assess how the current situation compares with the recent situation. This is useful for determining trends and the seriousness of the situation. Methods for estimating retrospective mortality include sample surveys and counting the number of recent graves.

Retrospective mortality survey

Death – the event being quantified – is relatively rare, and the sample therefore needs to be adequately large. For cluster sampling, 30 clusters of 30 families (approximately 4000 individuals) is an appropriate sample size.

The head of each family surveyed is asked about the occurrence (or not) of death within the household during a defined period of time. To avoid recall bias and to limit difficulties in interpreting the results, this period of time should be as short as possible but long enough to allow a sufficient number of "death events" to have occurred.

To calculate the mortality rate in the surveyed population:

- Divide the total number of deaths that have occurred in a sample (numerator) by the total number of surveyed individuals alive plus the number who have died (denominator) during a given period of time.
- Multiply the result by 10 000 and relate it to a period of one day to permit comparisons to be made.

For example: A survey of 5500 individuals showed that there were 49 deaths during the 28 days preceding the survey. The mortality rate was thus {[49/(5500 + 49)] x 10 000} / 28 = 3.2 deaths/10 000 population per day.

Counting recent graves

In some situations it is possible to count the number of graves dug since the arrival of the refugee or displaced population. Although this method is approximate, it provides some useful information when no other data are available.

For example: Six months after the arrival of a large number of displaced people at Hoddur in Somalia, 5900 graves were counted among a population of 25 000. Thus 5900/(5900 + 25 000) = 0.19 = 19% of the population died in 6 months, an average of 10.4 deaths/10 000 population per day.

ANNEX III

Overview of methods for conducting rapid assessments of malaria transmission[1]

Rapid assessments can be used to understand the level of malaria transmission in a given population. They can be carried out at health centre level or in the community.

In acute emergencies, when case loads are unusually high, clinic-based surveys can provide a very useful initial indication of the relative importance of malaria as a cause of morbidity at facility level. In well set-up camps or settlements, where health care is provided free of charge, it may be that everyone has access to health care (usually 5–10 times more than in "normal" conditions). In less well-organized situations, only a minority of the people who are actually sick may attend facilities and it may be difficult to ascertain how many people would usually attend under more normal circumstances, potentially resulting in a skewed picture of the malaria situation.

Data from health facilities only provide information about people who seek treatment at those facilities and therefore the data may not be representative of the overall population. Nevertheless, information from facilities may be used to give some idea of the level of transmission in the community if there is some understanding of the level of health care seeking behaviour and testing practices. If facility data is unavailable, a cross-sectional prevalence survey in the community may be necessary. The community survey may be conducted in different sites if there are reasons to suspect localized variations in malaria epidemiology.

Rapid clinic-based assessment

The purpose of a clinic-based assessment is to find out the prevalence of parasitaemia among symptomatic patients attending the clinic, i.e. the proportion of febrile patients with malaria parasites. If the clinicians at the facilities serving the population are following recommended diagnostic testing for malaria (whereby every suspected malaria case should be tested for malaria), clinic and laboratory records can be reviewed to assess the proportion of suspected malaria cases with confirmed malaria. If the diagnostic effort

[1] The methods described below are not intended to be a complete protocol for conducting a facility or community based survey of malaria.

at the facilities is uncertain, or if the emergency is at an early stage and diagnostic supplies have not yet been scaled up to recommended levels, the following survey approach can be used.

Conducting the survey

- Choose microscopy or RDTs. In the absence of a team of skilled microscopists, quality malaria RDTs give accurate, reliable and rapid results (< 30 minutes). Over a hundred RDTs are commercially available and vary by target species,[1] format (cassette, card, dipsticks) and performance (sensitivity and specificity). RDTs should be selected based on their performance in the WHO Malaria RDT Product Testing Programme and according to the WHO/GMP *Information note on recommended selection criteria for procurement of malaria rapid diagnostic tests (RDTs)*.[2] Furthermore, if species specific prevalence of *P. ovale*, *P. malariae* and *P. knowlesi* is relevant then microscopy is required.
- Microscopy through proper staining procedures and under supervision by skilled microscopists remains the gold standard for malaria diagnosis. Another advantage to microscopy is the fact that blood smears remain intact for months and can be checked (quality control assurance) by supervisors outside the population area. Whether microscopy or RDTs are selected performance is dependent upon end-users receiving training and regular supervision.
- Ensure there is sufficient staff to conduct the survey.
- Health facility or clinics should be selected to ensure they represent, as far as possible, the facilities accessed by the population. If it is impractical to include all facilities serving the population, a simple sampling procedure should be employed to select facilities.
- In a given facility, test all patients presenting with a febrile illness, using microscopy or RDTs, up to 100 patients. If possible do this in one day.
- Record the total number of outpatients seen at the clinic while the survey is being done.
- Record the age, sex and place of origin of all patients and the pregnancy status of female patients.
- Record the parasite rate or slide positivity rate (SPR) or RDT positive rate in patients with febrile disease or suspected malaria.

---

[1] *P. falciparum* only; *P. vivax* only; combination of *P. falciparum* and non-*P falciparum* species (pan, *P. vivax* or P.VOM-specific); or a single test line for all *Plasmodium* species (pan).

[2] http://www.who.int/malaria/publications/atoz/rdt_selection_criteria/en/index.html (accessed 16 April 2013)

- In the first few weeks of the acute phase, before diagnostic testing services have been established and a disease surveillance system has been put in place, rapid assessment surveys among febrile patients could be repeated each week to monitor changes in malaria transmission.

Table III.1 **Definitions**

Confirmed malaria case A suspected malaria case in which malaria parasites have been demonstrated by microscopy or RDT
Malaria test positivity rate (RDT and/or blood slide) The proportion of confirmed malaria cases among patients receiving a parasitological test (i.e. the proportion of RDTs and/or blood slides showing parasites among those tested).

Analysis and interpretation of results

- Record the total number of patients at the clinic, the total number of fever cases (suspected malaria), the proportion of fever cases among all outpatients, and the proportion of confirmed malaria cases among fever cases investigated (test positivity rate). Confirmed malaria cases and test positivity rates should be stratified by age <10 and >10 years of age, and by occurrence in host or displaced populations.
- Present the results graphically by week and age group.
- As mentioned above, facility data on malaria cases and test positivity rates can be used to assess the occurrence of malaria in the community; however, drawing conclusions about occurrence of malaria in the community from facility data should be done with an understanding of how facility data relates to occurrence of disease in the community.

In the host population

- If test positivity rates are less than 10% (parasites are found in fewer than 10% of fever cases) among both those under 10 years and those 10 years or older, this is an indication of low transmission (children and adults usually at risk of disease).
- If test positivity rates are greater than 50% in children aged under 10 years and less than 50% in those 10 years of age or older, this is an indication of high transmission (children most at risk).
- If test positivity rates are more than 50% in children aged less than 10 years and those 10 years or older, this is an indication of the possible onset of an outbreak.

In the displaced population

Analysis in the displaced population is more complex, since the same results may reflect different scenarios or situations (as listed below), especially in the first few weeks of settlement. A high proportion of fever cases testing positive for malaria in refugees or displaced people is of great concern and should be further investigated. It may be an indication of:

- High local malaria transmission:
  - If the incoming population is non-immune, high test positivity rates will be recorded in all age groups.
  - If the incoming population is semi-immune, children and pregnant women will have the highest test positivity rates.
- High malaria endemicity in the area of origin:
  - If there is low transmission in the resettlement area, test positivity rates among the displaced population may be very high initially (and greater in children than in adults) reflecting the level of transmission in their area of origin; however, lower test positivity rates will be observed over time as the occurrence of malaria disease in the displaced population reflects the lower transmission potential in the resettlement area.
- High malaria endemicity in areas the displaced population travelled through regardless of the level of transmission in the resettlement area:
  - If there is low transmission in the resettlement area, test positivity rates among the displaced population may be very high initially (and greater in children than in adults) reflecting the level of transmission in their area of origin; however, lower test positivity rates will be observed over time as the occurrence of malaria disease in the displaced population reflects the lower transmission potential in the resettlement area.
- The beginning of a malaria outbreak among the refugees or displaced population if they have come from a low to a high transmission area.

Rapid cross-sectional fever and parasitological prevalence survey across the affected population

The purpose of a community cross-sectional prevalence survey is to:

- provide pre-intervention baseline data;
- ascertain the prevalence of symptomatic and asymptomatic malaria infection (with confirmed parasitaemia) across all age groups;
- identify groups at risk and target interventions;

- gather information on the treatment-seeking behaviour of the population, for example, whether they received treatment in the previous 7 days, what treatment they received and where they received it.

Guiding principles

- Highly precise survey results are usually not needed during acute emergencies. The need is rather for acceptable precision in determining the level of malaria endemicity, particularly to distinguish between medium/high endemicity (more than 10% of people in the general population are infected) and low endemicity (well under 10% of people are infected), since the interventions to be planned are quite different.
- More precise survey data may be needed for operational research and for evaluating the impact of interventions over time and space.

In general, it can take 3–6 days to carry out a rapid cross-sectional malaria prevalence survey and it should be planned accordingly. It is useful to collect baseline information about haemoglobin levels where possible, because malaria is often an important cause of severe anaemia (Hb <5 g/dl) in endemic areas.

Conducting the survey

- *Define the study population.* If there are differences in malaria epidemiology, it will be necessary to define a study population within each type of ecological area or to take separate samples of the same size from each area.
- *Determine the sample size.* The required sample size depends on the expected prevalence of malaria illness and parasitaemia, on how precise the results need to be (i.e. how "wide" the 95% confidence interval can be allowed to be), on the sampling method (the "design effect"). For valid comparisons, a smaller sample is sufficient when the spleen or parasite rate is high in the investigated areas. For statistical validity of results, however, a large sample is required. The StatCalc program on EpiInfo, which can be downloaded free from http://www.cdc.gov/epiinfo, can be used to help in the calculation of sample size.

For example: The prevalence is assumed to be 50%; 45% is used as the "lowest acceptable result"; the population is 25 000; the chosen confidence interval is 95%. The desirable sample size using EpiInfo would be 378. With an assumed design effect of 2 (depending on the sampling method), the required sample size is 2 x 378 = 756.

- Confidence interval. A 95% confidence interval is a commonly used index of statistical variability when analysing survey data. It can be roughly interpreted to mean that there is a 95% chance that the true values will fall within this interval. The larger the sample size, the more accurate is the measured prevalence rate expected to be, and the narrower the 95% confidence interval. Survey reports should preferably provide a 95% confidence interval in the results. Confidence intervals can become fairly wide when data need to be broken down by age – which is probably adequate for purposes of planning malaria control programmes. However, to demonstrate a statistically significant change in prevalence associated with a particular intervention, the size of the effect would need to be very large to be detected. If the results from the study are to show that an intervention has been effective, a larger sample size may be needed.

For example: If in a sample of 100 children parasites are found in 50, the parasite rate in the total population of children of the same age group lies between 40% and 60%, and this may be assumed with a 95% probability. With the same parasite rate result (50%) in a sample of 500 children, the 95% confidence interval would be 46–54% (Bruce-Chwatt, 1985). See Table III.2

Note: These figures are for random sampling, *not* for cluster sampling. Random sampling selects sample units such that each possible unit has a fixed and determinate probability of selection (Last, 2001). Thus, each person has an equal chance of being selected out of the total study population. Random sampling should not be confused with haphazard selection of study subjects. Random sampling follows a predetermined plan that is usually devised with the aid of a table of random numbers.

Table III.2 **Confidence interval at 95% probability level for a standardized sample of 100 and 500, using random sampling**

Prevalence	60	50	40	10	5
95% confidence interval, with a sample size at random of 100	50–70	40–60	30–50	5–18	2–11
95% confidence interval, with a sample size at random of 500	56–64	46–54	36–44	8–13	3–7

- *Choose a sampling method.* There are many sampling methods. For malaria, everyone in the household is usually surveyed. Each household must have an equal chance of being selected. Methods of selecting households include random sampling, systematic sampling, and random cluster sampling. However, in the acute phase of emergencies, households may not be stable or well defined, and the most appropriate method in the first few weeks of this phase may be random sampling.

- *Recruit and train survey team, selecting one person as team leader.* The team should include:
  - — one person to explain the procedure and obtain informed consent from the household. Ideally this person should be a community member;
  - — one person to register the participants, record the axillary temperature the RDT result, and information on treatment-seeking behaviour;
  - — one person to take the axillary temperature and perform the RDT or make a malaria smear on a slide;
  - — one person to measure spleen size (if this is to be done) or, more feasibly in emergencies, to check for the presence of enlarged spleen;
  - — one person to administer treatment (if this is given on the spot) and to provide advice about danger signs and follow-up.
- *Ensure adequate supplies.* The following are essential:
  - — rapid tests or microscope slides and associated consumables (e.g. lancets, rubber gloves, alcohol for swabbing skin, cotton wool, suitable disposal facilities for sharps and contaminated materials);
  - — referral forms for the nearest suitable clinic;
  - — administrative equipment, e.g. log book, consent forms, record forms, pens, pencils and markers;
  - — funds to pay for transport to the clinic if needed.
- *Repeat the survey.* If malaria transmission is seasonal, it can be useful to repeat the survey in different seasons and compare the results. For example, in some areas prevalence is lower at the end of the dry season than at the end of the wet season. In areas where malaria transmission is intense all year round, repeating the survey several months after the start of the emergency response can provide a good indication of the impact of interventions. In areas with marked seasonal peaks it may be useful to repeat the survey after 12 months (i.e. at the same time of year) to measure any impact of interventions.

Analysis and use of results

- *Prevalence of malaria.* Calculate the proportion of people, out of the entire population, with fever or a history of fever who have a positive RDT (or slide) result. This will provide an estimate of the prevalence of malaria in the study population at that specific point in time. Analyse the data by sex and by age group. The number of pregnant women in the sample will probably be too low to be statistically significant.

- *Endemicity.* Calculate the proportion of people who have a positive RDT or slide result. Use the parasite rates to estimate the underlying endemicity of malaria (see Table III.3).

Table III.3 **Malaria endemicity according to spleen and parasite rates**

Malaria endemicity	**Parasite rate (PR)**	
	Under 5 years	**Over 5 years**
Meso and hyper-endemic (usually linked to transmission)	High (>50%), but lower than spleen rate	Low
Hypoendemic (usually linked to low transmission	≤10%, always higher than spleen rate	

In cross-sectional surveys, parasite and spleen rates are normally determined in children aged 2–9 years. In areas where malaria transmission is high, parasite and spleen rates are highest in children aged under 5 years and decline above that age. In areas where malaria transmission is low, parasite and spleen rates are low and are equally distributed over all age groups.

- *Representativeness of the sample.* Check the survey data to see whether there are many absentees, whether the sex and age of absentees are statistically different from the group sampled, and whether particular groups (e.g. children aged under 5 years) are underrepresented. Also review the sex and age groups of the sample to see whether the proportions are similar to those in the whole community.
- Enter the data into a software programme to help with analysis.

References

- Bruce-Chwatt (1985). *Essential Malariology*: Second Edition: London: William Heineman, xii + 452 pp., illustrations.
- Last J (2001). *A dictionary of epidemiology*, 4th ed. Oxford, Oxford University Press.

ANNEX IV

Example malaria rapid prevalence data collection tool

Malaria prevalence survey form Date : _ _ / _ _ / _ _ _ _ Code _ _ _ _ _ _ _ _
Investigator's name: Household number: _ _ _ _
Health zone:
Village:

Serial number	Sex		Age	History of fever during the past 48 hours?		Axillary temp-erature	Rapid test *P. falciparum* (RDT)		Treatment received in the past 7 days?		Where was treatment sought?						Treatment received?					Comments
	M	F	Years (<1 year = 0)	Yes	No	° C	+	–	Yes	No	hf	pf	m	ps	th	s	cq	sp	q	ACT	o	

Legend
hf: health facility pf: private health facility m: market ps: private shop th: traditional healers s: personal stock
cq: chloroquine sp: sulfadoxine–pyrimethamine q: quinine ACT: artemisinin-based combination therapy o: other

ANNEX V

Checklist for effective response to malaria epidemics

Preparedness and early detection

Pre-season preparedness and early identification provide the malaria manager with an increasing number of tools to deal with an epidemic. Maintain a high-quality surveillance system, keep databases up-to-date, make and analyse regular reports. Think ahead – **be prepared.**

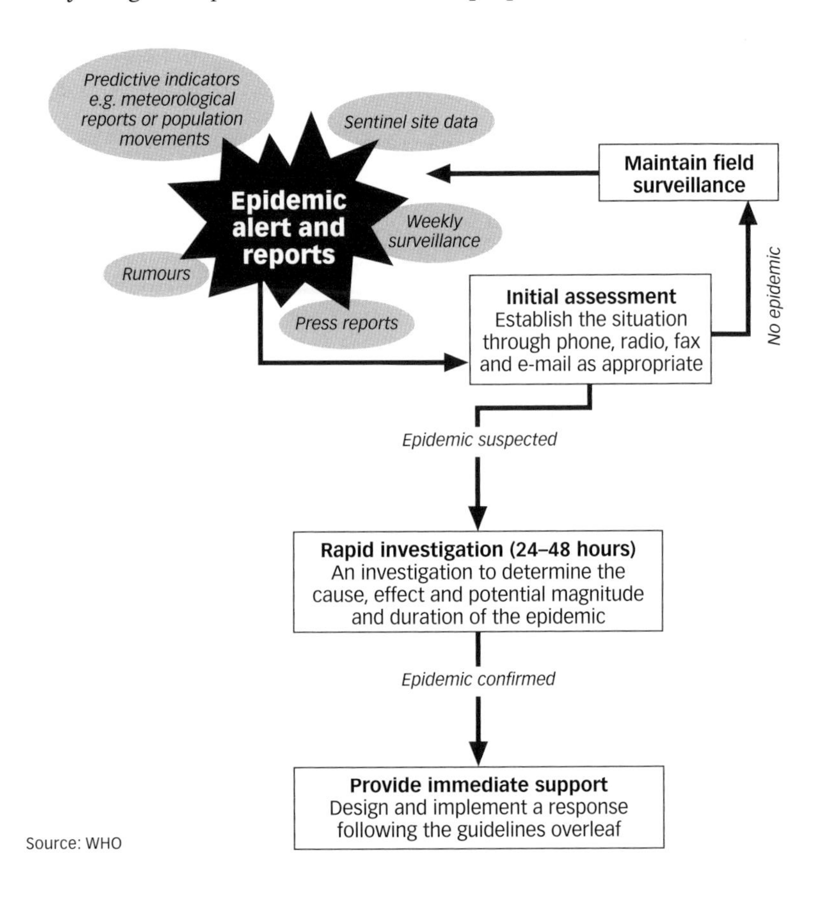

Source: WHO

Designing the response

At all epidemic stages

- Ensure that all clinics and health facilities are operational and have sufficient drugs, equipment and trained staff;
- Establish treatment centres (e.g. temporary clinics, mobile clinics) in areas where access is a problem or health facility coverage is low;
- Ensure that appropriate diagnostic testing and treatment is provided at all health facilities and at community level;
- Promote outreach activities for early identification of malaria patients in the community, and organize necessary case management/referral;
- Reinforce the referral system and consider the introduction of artesunate suppositories or IM artemether as pre-referral treatment for severe malaria where these are not already used;
- Intensify/maintain effective preventive measures for all at risk populations;
- Reinforce health information systems for reporting and epidemic monitoring, preferably on a weekly basis;
- Conduct specific epidemic health education campaigns;
- Organize regular press releases, conferences, and articles for public information.

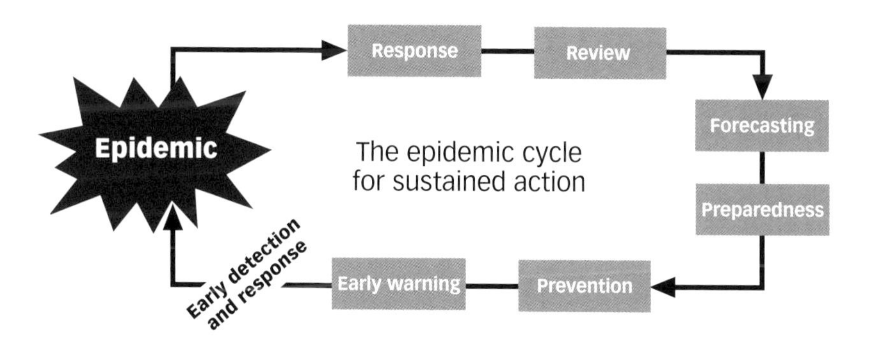

Epidemic start

In the early stages of an epidemic, in addition to reinforcing case management, it is still possible to reduce the potential impact of the epidemic using vector control interventions, such as IRS and/or LLINs (see Chapter 7).

Impact

Time

Additional specific interventions include:

- *if the area is already protected by IRS*, establish coverage and quality of vector control using techniques such as bioassays;
- *if the area is not already protected by IRS*, but the malaria epidemiology, type of housing, and available logistics would allow rapid deployment of effective IRS, implement IRS in target areas;
- consider distribution of LLINs if there is a history of LLIN use in the area and/or if the capacity exists to implement such a programme.

Epidemic acceleration

Case management (see above) is the priority at this stage. Aiming to reduce the potential impact of the epidemic through IRS and other vector control methods is an option only if (i) there is considerable operational capacity trained and readily available, and (ii) preparedness levels have been high and sufficient resources are available for high coverage.

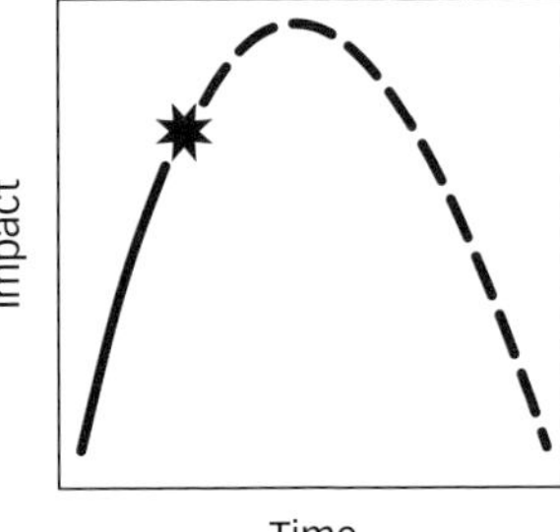

Additional specific interventions to consider include:

- establishing coverage and quality of vector control, using techniques such as bioassays, if the area has already been sprayed;
- IRS if the area is not already protected;
- introduce rotational use of insecticides for IRS even when observed vector susceptibility is high;
- properly-timed fogging in highly populated areas such as refugee/displaced persons camps, especially if shelters are small and IRS is not an option.

Epidemic peak

The epidemic has already begun to stabilize at this point and the number of new cases per day or week is stabilizing or beginning to decrease. Vector control interventions to reduce the potential impact of the epidemic will have limited public health value at this late stage. Resources should instead be directed at case management to reduce malaria mortality (see above).

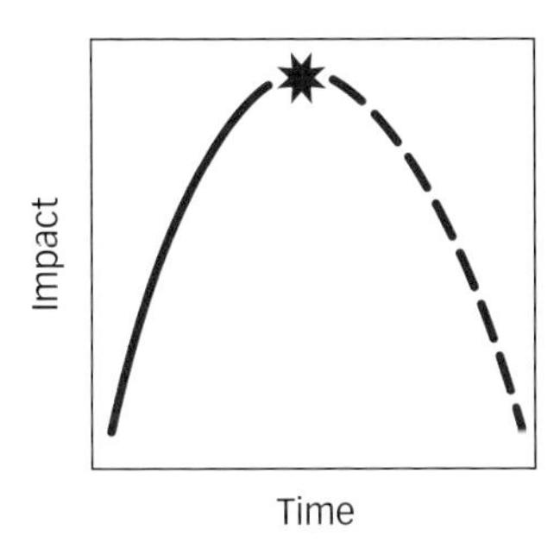

Post-epidemic

Ensure that there is a lessons-learnt exercise that examines the programme and how it has performed. This will improve response in subsequent epidemics.

ANNEX VI

Antimalarial treatment regimes

Treatment regimens for uncomplicated malaria

ACT options recommended by WHO are:

- Artemether plus lumefantrine (AL)
- Artesunate plus amodiaquine (AS+AQ)
- Dihydroartemisinin plus piperaquine (DHA+PPQ)
- Artesunate plus mefloquine (AS+MQ)
- Artesunate plus sulfadoxine-pyramethamine (AS+SP)

Artemether plus lumefantrine (AL)

Presentation: Artemether plus lumefantrine (AL) tablets 20 mg + 120 mg, also available as dispersible tablets.

AL is available as a prequalified fixed dose combination from a number of manufacturers. AL is particularly useful in humanitarian emergencies where there are no local data on the therapeutic efficacy of antimalarial medicines, as it provides high cure rates against falciparum malaria in most settings.

The absorption of lumefantrine, a hydrophobic, lipophilic compound, is variable and increases when the drug is taken with food, particularly fatty foods. Drug absorption therefore increases as the appetite returns with clinical recovery. Absorption is increased by more than 100% when AL is taken with a meal rich in fat. Patients should be advised to take the medication with food, and, if possible, the patient should be provided with milk or a similar drink to take with the tablets, particularly in the second and third day of treatment. AL needs to be taken twice a day for three days and in adults this requires four tablets per dose, up to a total of 24 tablets.

Side-effects: Although AL is generally well tolerated, reported side-effects include dizziness and fatigue, anorexia, nausea, vomiting, abdominal pain, palpitations, myalgia, sleep disorders, arthralgia, headache, and rash. Side-effects are generally mild and there is no indication of cardiotoxicity.

Contraindications:

— Persons with known hypersensitivity to either component drug;
— Severe malaria according to WHO definition.

Safety in special groups. The safety of AL has not yet been established in children of less than 5 kg body-weight, or in the first trimester of pregnancy. Its use in these groups is not recommended until more safety data becomes available. However if there are no other effective options available, it can be used under medical supervision.

Weight/Age	Number of tablets and approximate dosing times					
	0 hours	8 hours	24 hours	36 hours	48 hours	60 hours
5–14.9 kg <3 years	1	1	1	1	1	1
15–24.9kg 3–8 years	2	2	2	2	2	2
25–34.9 kg 9–13 years	3	3	3	3	3	3
>35 kg >14 years	4	4	4	4	4	4

To simplify the schedule:

- The 2nd dose can be given anytime from 8–12 hours after the first dose;
- Thereafter, doses can be given twice a day morning and evening.

Artesunate plus Amodiaquine (AS+AQ)

Presentation: AS + AQ is available as FDC tablets in three strengths: 67.5 mg + 25 mg, 135 mg + 50 mg, and 270 mg +100 mg. AS+AQ is available as a prequalified fixed dose combination from a number of manufacturers.

Several co-blistered formulations are also available on the market, but wherever possible the fixed-dose combination should be used as preferred treatment to improve patience adherence to treatment.

It is not necessary to take the tablets with food and the number of tablets per dose is low: 1 tablet once daily in infants, children and adolescents and two tablets once daily for adults. Infant tablets can be dissolved in a small amount of liquid making them easy to administer. Efficacy of AS+AQ remains acceptably high in most settings of Western and Central sub-Saharan Africa.

Side-effects: Those reported include nausea, vomiting, abdominal pain, diarrhoea, itching, and less commonly, bradycardia.

Contraindications:
— Persons with known hypersensitivity to either component drug;
— Severe malaria according to WHO definition;
— Persons with hepatic disorders;
— HIV-infected patients receiving zidovudine or efavirenz.

Safety in special groups. There is no data on the safety of AS+AQ in young children under 4.5 kg or during the first trimester of pregnancy. Its use in these groups is not recommended until more safety data becomes available.

Weight/Age	AS+AQ FDC (25 mg + 67.5 mg)			AS+AQ FDC (50 mg + 135 mg)			AS+AQ FDC (100 mg + 270 mg)		
	Day 1	Day 2	Day 3	Day 1	Day 2	Day 3	Day 1	Day 2	Day 3
≥4.5 to <9 kg 2–11 months	1	1	1						
≥9 to <18 kg 1–5 years				1	1	1			
≥18 to <36 kg 6–13 years							1	1	1
≥36 kg 14+ years							2	2	2

Dihydroartemisinin plus Piperaquine (DHA+PPQ)

Presentation: Dihydroartemisinin (DHA) plus Piperaquine (PPQ) is available as FDC tablets in two tablet strengths: 20 mg + 160 mg and 40 mg + 320 mg. DHA + PPQ is generally well tolerated. In many countries, AL or AS + MQ, or DHA + PPQ may give the highest cure rates. A dose of DHA + PPQ should be administered once daily (at the same time each day) over three consecutive days for a total of three doses. Each dose should be taken with water and without food.

Side effects: Reports include nausea, vomiting, abdominal pain, gastrointestinal effects and dizziness. No serious adverse events have been reported.

Contraindications:
— Persons with known hypersensitivity to either component drug;
— Severe malaria according to WHO definition

Safety in special groups. There is no data on the safety of DHA + PPQ in young children under 5 kg or during the first trimester of pregnancy. Its use in these groups is not recommended until more safety data becomes available.

Weight (kg)	DHA+PPQ FDC (20 mg + 160 mg)			DHA+PPQ FDC (40 mg + 320 mg)		
	Day 1	Day 2	Day 3	Day 1	Day 2	Day 3
5 to <7	½	½	½			
7 to <13	1	1	1			
13 to <24				1	1	1
24 to <36				2	2	2
36 to <75				3	3	3
75 to < 100				4	4	4

Artesunate plus Mefloquine (AS+MQ)

Presentation: Artesunate (AS) plus Mefloquine (MQ) is available as FDC tablets in two tablet strengths: 25 mg + 50 mg (mefloquine as hydrochloride) and 100 mg + 200 mg. Artesunate plus mefloquine is available as a prequalified fixed dose combination from a single manufacturer. Several co-blistered formulations are also available on the market, but wherever possible the fixed-dose combination should be used as the preferred treatment to improve patience adherence to treatment. As with AL and with DHA + PPQ, artesunate plus mefloquine is suitable for emergencies in areas where the efficacy of SP and amodiaquine is unknown.

Delivery: Vomiting especially in young children can be reduced if paracetamol is given ½ hour before administration of the first dose. Mefloquine absorption is improved when the drug is taken with plenty of water or following food. AS + MQ should not be used in re-treatment within eight weeks of previous treatment.

Side-effects: The most common side-effects for mefloquine are nausea, vomiting, dizziness, sleep disorders, anxiety and neurological symptoms. Serious adverse neuropsychiatric reactions have been reported in people taking mefloquine, including psychosis, toxic encephalopathy, convulsions and acute brain syndrome. The frequency of serious side-effects appears to be dose-related, with reported risk 7-fold higher among people who have been re-treated with mefloquine within 4 weeks of previous treatment.

Contraindications:

— Persons with known hypersensitivity to either component drug;
— Severe malaria according to WHO definition;
— Persons with history of severe neuropsychiatric disease;
— Persons who have received mefloquine in the previous 8 weeks;

Safety in special groups. There is no data on the safety of AS + MQ in young children under 5 kg or during the first trimester of pregnancy, or in patients weighing more than 69 kg. Its use in these groups is not recommended until more safety data becomes available.

Weight/Age	**AS+MQ FDC (25 mg + 50 mg)**			**AS+MQ FDC (100 mg + 200 mg)**		
	Day 1	**Day 2**	**Day 3**	**Day 1**	**Day 2**	**Day 3**
≥5 to 8 kg 6–11 months	1	1	1			
≥9 to <18 kg 1–5 years	2	2	2			
≥18 to <30 kg 6–13 years				1	1	1
≥30 kg 14+ years				2	2	2

Artesunate plus sulfadoxine–pyrimethamine (AS+SP)

Presentation: Artesunate (AS) tablets in two strengths (50 mg and 100 mg tablets) plus sulfadoxine/pyrimethamine (SP) tablets 25 mg + 500 mg, only available as co-blistered formulation.

This combination is an option for use in emergencies in countries where it is the recommended 1st for treatment of malaria. Patients should not be treated with this ACT if they have received SP in the previous 4 weeks or if they have known hypersensitivity to sulfa drugs. While cutaneous reactions are quite rare, they are more common in PLHIV and this should be taken into account when dealing with known high-risk groups, especially during the chronic phase of the emergency where sexual abuse and sex work may be prevalent.

Delivery: The practical operational advantage of SP over other co-blistered ACTs is that the full dose of SP is delivered under directly observed treatment as a single dose on day 1, together with the first dose of artesunate. SP is less likely than mefloquine to cause vomiting in young children.

Side-effects. SP is generally well tolerated when used as recommended and severe adverse side-effects are rare. However, allergic reactions of varying severity have been reported and are more common after repeated dosing.

Contraindications:
— persons with known hypersensitivity to either component drug;
— severe malaria according to WHO definition;
— PLHIV receiving co-trimoxazole prophylaxis;
— pregnant women who received SP for intermittent preventive treatment (IPTp) within the previous 4 weeks;
— patients with severe hepatic or renal dysfunction.

Safety in special groups. There is no data on the safety of AS + SP in young children under 5 kg, during the first trimester of pregnancy, or in patients weighing more than 69 kg. Its use in these groups is not recommended until more safety data becomes available.

Weight / Age	No of AS tablets (50 mg)			No of AS tablets (100 mg)			No of SP tablets (25 mg + 500 mg)		
	Day 1	Day2	Day 3	Day 1	Day2	Day 3	Day1	Day 2	Day 3
<10 kg 5–11 months	½	½	½				½		
≥10 to <21 kg 1–6 years	1	1	1				1		
≥21 to <37 kg 7–13 years				1	1	1	2		
≥ 37 kg 14 years				2	2	2	3		

Quinine

Presentation: Tablets of quinine hydrochloride, quinine dihydrochloride and quinine sulfate containing 82%, 82%, 82.6% quinine base respectively. Generally available as quinine sulfate or hydrochloride tablets of 200 mg and 300 mg respectively, but can also be found in 100 mg and 500 mg tablets.

Delivery: Quinine should be given as 30 mg salt/kg/day in three divided doses at 8 hourly intervals for seven days and should be combined with **doxycycline** 100 mg twice a day for seven days. For pregnant women and children under 8 years of age, whereby doxycycline is contraindicated, quinine should be given combined with **clindamycin** 7–13 mg/kg every

8 hours for five days. Quinine with clindamycin is the treatment of choice for uncomplicated falciparum malaria during the 1st trimester. Adherence to treatment with quinine is problematic, particularly because of its poor tolerability.

Side-effects. Administration of quinine causes complex symptoms known as cinchonism, which is characterized in its mild form by tinnitus, impaired high tone hearing, headache, nausea, dizziness and dysphoria, and sometimes disturbed vision. More severe manifestations include vomiting, abdominal pain, diarrhoea and severe vertigo. The most common adverse effect in the treatment of severe malaria is hyperinsulinaemic hypoglycaemia, particularly in late pregnancy.

Contraindications:
- Persons with known hypersensitivity;
- Drugs that may prolong the QT interval should not be given with quinine.

Weight (kg)	Quinine salt 100 mg tabs (x 3 = daily dose)	Quinine salt 200 mg tabs (x 3 = daily dose)	Quinine salt 300 mg tabs (x 3 = daily dose)
3–6	½		
7–12	1	½	
13–17	1½		½
18–25	2	1	
26–35		1½	1
36–50		2	
>50		3	2

Chloroquine

Presentations: Available as chloroquine sulfate or phosphate 100 mg base tablets,150 mg base tablets and 50 mg base/5 ml syrup.

Chloroquine should only be used for the treatment of chloroquine-sensitive uncomplicated non-falciparum malaria, i.e. *P. vivax*, *P. malariae* and *P. ovale*. *P. knowlesi* is also sensitive to chloroquine. Only in Central America does *P. falciparum* remain sensitive to chloroquine. Chloroquine has a low safety margin and overdosage is very dangerous, particularly when the drug is being used for the treatment of rheumatoid arthritis.

Dose: Total dose is 25 mg/kg given over three days (up to a total dose of 1500 mg base for adults)

Side-effects: At the doses used for the treatment of malaria, chloroquine is generally well tolerated. The key adverse side-effects are the unpleasant taste of the drug, which may upset children, and pruritus, which may be severe in dark-skinned patients. Other less common side effects include headache, various skin eruptions and gastrointestinal disturbances, such as nausea, vomiting and diarrhoea.

Contraindications:

— Persons with known hypersensitivity;

— Drugs that may prolong the QT interval should not be given with chloroquine.

Weight (kg)	Age	Chloroquine base 100 mg tabs			Chloroquine base 150 mg tabs		
		Day 1	Day 2	Day 3	Day 1	Day 2	Day 3
5–6	<4 months	½	½	½	½	¼	¼
7–10	4–11 months	1	1	½	½	½	½
11–14	1–2 years	1 ½	1 ½	½	1	1	½
15–18	3–4 years	2	2	½	1	1	1
19–24	5–7 years	2 ½	2 ½	1	1 ½	1 ½	1
25–35	8–10 years	3 ½	3 ½	2	2 ½	2 ½	1
36–50	11–13 years	5	5	2 ½	3	3	2
>50	>14	6	6	3	4	4	2

Other recommended regimes:

— Initial dose: 10 mg base /kg (max 600 mg)

— Followed by: 5 mg base/ kg (max 300 mg) at 6–8, 24 and 48 hours

Primaquine

Presentations: Primaquine (PQ) is available as tablets containing 5 mg, 7.5 mg or 15 mg tablets as diphosphate.

Primaquine is highly active against mature gametocytes (the sexual form of the parasite which transmits malaria) and against the hypnozoites of the relapsing malaria species, *P. vivax* and *P. ovale*.

Dose: A single dose of 0.25 mg base/kg of primaquine is an effective gametocytocide for *P. falciparum*. The standard anti-relapse treatment for the relapsing malaria species, *P. vivax* and *P. ovale*, is 0.5 mg primaquine base/kg per day for 14 days in South-east Asia and Oceania, and 0.25 mg primaquine base/kg per day for 14 days in other areas.

Side-effects: The most important adverse effects are haemolytic anaemia in patients with G6PD deficiency. In patients with the African variant of G6PD deficiency, the standard course of primaquine generally produces a benign self-limiting anaemia. In the Mediterranean and Asian variants, haemolysis may be severe. Therapeutic doses may also cause abdominal pain if administered on an empty stomach. Larger doses can cause nausea and vomiting. Methaemoglobinaemia may occur.

Contraindications:

- Persons with known hypersensitivity;
- Persons with G6PD deficiency
- Pregnant women and infants below 1 year of age.

Resistance to primaquine is emerging in the Western Pacific region, Southeast Asia, South America and parts of Africa. Relapses occurring in patients taking the standard regime should be treated with an increased dose of 30 mg/day for 14 days.

Where ACT is first-line treatment for *P. falciparum*, it can be used for *P. vivax* in combination with PQ for radical cure – but AS + SP will not be effective against *P. vivax* in many places.[1]

Treatment regimens for severe malaria

Artesunate

Presentations: Artesunate is available in vials containing artesunate powder for injection, containing artesunate 30 mg, 60 mg and 120 mg, co-packed with 1 ampoule of 5% sodium bicarbonate injection (1 ml) and 1 ampoule of sodium chloride injection (5 ml).

Intravenous or intramuscular artesunate should be used in preference to parenteral quinine for the treatment of severe *P. falciparum* malaria, as it reduces significantly malaria mortality; is associated with a lower risk of hypoglycaemia; and does not differ in terms of risk of serious neurological sequelae. Moreover, artesunate offers a number of programmatic advantages over quinine as it does not require rate-controlled infusion or cardiac monitoring.

Dose: Artesunate should be given at 2.4 mg/kg by direct IV (slow bolus, 3–4 ml per minute) or IM injection at 0, 12 and 24 hours; then once daily until oral therapy is possible. As soon as the patient can swallow, a full ACT treatment course should be given.

---

[1] Recommend either of these regimens for anyone who was infected with *P. vivax* in Papua New Guinea or Indonesia.

Delivery: The vial of artesunate powder should be mixed with 1 ml of 5% sodium bicarbonate solution (provided) and shaken for 2–3 minutes for better dissolution. Then,

1. for IV administration: add 5 ml of 5% glucose or normal saline to make the concentration of artesunate as 10 mg/ml and administer by slow bolus infusion; OR
2. for IM administration: add 2 ml of 5% glucose or normal saline to make the concentration of artesunate as 20 mg/ml

The correct dose in ml of artesunate should be calculated on the basis of the patient's body weight.

- The required dose in ml of artesunate solution for IV administration = 2.4 x body weight (kg)/10
- The required dose in ml of artesunate solution for IV administration = 2.4 x body weight (kg)/20

Artemether

Presentations: Available in ampoules containing artemether injectable solution for intramuscular injection containing 100 mg or 80 mg of artemether in 1 ml for adults, and 20 mg of artemether in 1 ml for paediatric use.

Intramuscular artemether or intravenous quinine should only be given if artesunate is not available.

Dose: IM: loading dose 3.2 mg/kg on Day 0, followed by a maintenance dose of 1.6 mg/kg daily for at least 2 more doses. As soon as the patient can swallow provide a full ACT treatment course.

Delivery: Artemether should be administered in the antero-lateral thigh. The 20 mg/ml presentation and a 1ml syringe facilitates dosing in small children. Artemether is not well absorbed in shock; an alternative treatment (parenteral or rectal artesunate, IV quinine) should be chosen.

The correct dose in ml of artemether should be calculated on the basis of the patient's body weight.

- The required loading dose in mg of artemether for IM administration = 3.2 x body weight (kg)
- The required maintenance dose in mg of artemether for IM administration = 1.6 x body weight (kg)

Quinine

Presentation: Injectable solutions of quinine hydrochloride, quinine dihydrochloride and quinine sulfate containing 82%, 82% and 82.6% quinine base respectively. Intravenous quinine or intramuscular artemether should only be given if artesunate is not available.

Dose: A loading dose of 20 mg/kg salt, then 10 mg/kg salt every 8 hours. The correct dose in mg of quinine should be calculated on the basis of the patient's body weight.

- The required loading dose in mg of quinine is = 20 mg quinine salt x body weight (kg)
- The required maintenance dose in mg of quinine is = 10 mg quinine salt x body weight (kg)

> **Caution:** Omit the loading dose if the patient has received 3 or more doses of quinine in the last 48 hours, or mefloquine or halofantrine within the last 3 days.

Delivery: The dose required is diluted with 5 or 10% glucose IV solution. Quinine is administered by rate-controlled low infusion over 4–8 hours, and the infusion rate should not exceed 5 mg salt/kg body weight per hour. In **children under 20 kg**: in a volume of 10 ml/kg infused over 4 hours; or 20 ml/kg infused over 8 hours. In **adults**: in a volume of 250 ml infused over 4 hours; or 500 ml infused over 8 hours. Monitor for hypoglycaemia and fluid overload, especially in children. Fluid used to administer quinine should be included in calculated fluid requirements. Quinine may also be given intramuscularly into the anterior thigh (not the buttock) after dilution to 60–100 mg/ml.

When the patient has received at least 3 parenteral doses of quinine, and is able to swallow, switch to oral administration: a 3 day course of **ACT** (If the combination AS+MQ is used, wait 12 hours after the last dose of quinine before giving MQ. Do not use AS+MQ if the patient developed neurological signs during the acute phase).

Rectal Artesunate

Presentation: Rectal capsules containing artesunate 50 mg and 200 mg

Recommended as pre-referral treatment of children with severe malaria where parenteral therapy with artesunate or quinine is not available or feasible. Referral to a health centre or hospital is indicated for young children who cannot swallow antimalarial medicines reliably. If referral is expected to take more than six hours, pre-referral treatment with rectal artesunate is indicated.

Dose: 10 mg/kg. The correct dose of rectal artesunate should be calculated on the basis of the patient's body weight.

Management of complications of severe malaria

Convulsions	Treatment
	• maintain the airway; • turn the patient on his or her side to reduce the risk of aspiration; • do not attempt to force anything into the patient's mouth; • check blood glucose and treat if <2.2 mmol/l (see below); • treat with: — diazepam, 0.3 mg/kg (up to maximum 10 mg), as a slow IV injection over 2 minutes; ***or*** — diazepam, 0.5 mg/kg, administered rectally by inserting a 1 ml syringe (without a needle); • Diazepam may be repeated if seizure activity does not stop after 10min. Midazolam may be used (same dose) instead of diazepam by either the intravenous or buccal route; • If the patient continues to convulse after 2 doses of diazepam, give phenytoin (18 mg/kg loading dose and then a maintenance dose of 5 mg/kg/day for 48 hours); • If these are not available or fail to control seizures, phenobarbitone may be used (15 mg/kg loading dose IM or slow IV infusion, then a maintenance dose of 5 mg/kg/day for 48 hours). When phenobarbitone is used, monitor the patient's breathing carefully as ventilatory support may be needed. High dose (20 mg/kg) phenobarbitone may lead to respiratory depression and an increased risk of death. Be prepared to use 'bag-and-mask' manual ventilation if the patient breathes inadequately.

Hypoglycaemia	Treatment
Unconscious patients with threshold glycaemia below 3 mmol/l (threshold for intervention)	Insert an IV line: • *Children:* Correct hypoglycaemia with 200–500 mg/kg of glucose. Immediately give 5 ml/kg of 10% dextrose through a peripheral line, followed by a slow intravenous infusion of 5 ml/kg/h. If 10% dextrose is not available dilute each 0.4 ml/kg of 50% dextrose with 1.6 ml/kg infusion fluid to form a 10% solution (e.g. for a 10 kg baby, dilute 4 ml of 50% with 16 ml of infusion fluid). Administration of hypertonic glucose (> 20%) is not recommended, as it is an irritant to peripheral veins. • *Adults:* give 25 g of dextrose preferably as 10% dextrose over a few minutes. If 10% dextrose is not available, 50ml of 50% dextrose (25 g) diluted with 100 ml of infusion fluid may be used and infused over a period of about 3–5 minutes. Follow with an intravenous infusion of 200–500 mg/kg per hour of 5% or 10% dextrose. • Recheck blood glucose with a rapid "stix" method 15 minutes after the end of the infusion. • If blood glucose is still <2.2 mmol/l, repeat glucose infusion as above. • If it is not possible to insert an IV line and the patient is unconscious, give 1 ml/kg 50% dextrose via nasogastric tube • Give oral fluids (breast milk or sugar solution) and food once the patient regains consciousness.
Unconscious patient, suspected clinically but not possible to check the blood glucose	• Give a presumptive intravenous infusion as described above. *Note:* There is a marked risk of hypoglycaemia with the use of quinine in pregnant women

Severe anaemia	Treatment
Children	• Generally, in high transmission settings EVF of ≤12% or a haemoglobin concentration of ≤4 g/dl is an indication for blood transfusion, whatever the clinical condition of the child. In low-transmission settings a threshold of 20% haematocrit or a haemoglobin of 7g/dl, is recommended for blood transfusion: — Give 10 ml of packed cells or 20 ml of screened, compatible whole blood per kg of body weight over a 4-hour period. • For children with less severe anaemia (i.e. EVF 13–18%, Hb 4–6 g/dl), transfusion should be considered for high-risk patients with any one of the following clinical features: (i) respiratory distress (acidosis); (ii) impaired consciousness; (iii) hyperparasitaemia, >20%; (iv) shock; (v) heart failure. • For children with severe anaemia and severe malnutrition (i.e. severe wasting plus oedema), give blood much more cautiously: — Infuse 10 ml/kg blood over 3 hours. — Give furosemide, 1 mg/kg IV, halfway through the transfusion to avoid circulatory overload. • Follow-up of haemoglobin (haematocrit) levels after blood transfusion is essential, as many children need a further transfusion within the next few hours, days or weeks.
Adults	• Anaemia is common in severe malaria and may be associated with secondary bacterial infection. Blood transfusion is indicated in patients with EVF of ≤20% or with haemoglobin concentration <7 g/dl: — 500 ml of screened, compatible fresh blood should be transfused over 6 hours. — Give a small intravenous dose of furosemide, 20 mg, during transfusion to avoid circulatory overload.

Shock	Treatment
Children	Common causes of shock include dehydration from diarrhoea, sepsis, anaemia. Classify the child as having SHOCK if the systolic blood pressure is < 50 mm Hg, pulse is weak and rapid (other signs include cold hands and/or feet AND capillary refill longer than 2 seconds). Shock in children with severe malaria is rare (<2% of cases of severe malaria). *Rapid administration of IV fluids for resuscitation of children in shock or with severe dehydration.* • Children with severe dehydration should be given rapid IV rehydration followed by oral rehydration therapy. The best IV fluid solution is Ringer's lactate solution (also called Hartmann's solution for injection). If Ringer's lactate is not available, normal saline solution (0.9% NaCl) can be used. 5% glucose (dextrose) solution on its own is not effective and should not be used. • Rapid fluid boluses are contraindicated in severe malaria resuscitation. • Give 100ml/kg of the chosen solution as follows: In children <12 months old give 30 ml/kg body weight in 1 hour, then 70 ml/kg bw over the next 5 hours. While in children ≥12 months old, give 30 ml/kg over 30 mins, then 70 ml/kg over next 2 hours. Repeat the first dose of 30 ml/kg if the radial pulse is still very weak or not detectable. — Correct hypovolaemia with maintenance fluids at 3–4 ml/kg per hour. — Keep an accurate record of fluid intake and output. Remember to include the volume of transfused cells or blood in calculating fluid balance. Monitor blood pressure, urine volume (every hour) and jugular venous pressure. — Improve oxygenation by clearing the airway, increasing the concentration of inspired oxygen and supporting ventilation artificially, if necessary.

<table>
<tr><td>Adults</td><td>Some patients are admitted in a state of collapse, with a systolic blood pressure < 80 mm Hg in the supine position; cold, clammy, cyanotic skin; constricted peripheral veins; and a rapid, feeble pulse. This clinical picture may indicate complicating septicaemia, and possible sites of associated bacterial infection should be sought, e.g. meningitis, pneumonia, urinary tract infection (especially if there is an indwelling catheter) or intravenous injection site infection.

Rapid administration of IV fluids for resuscitation of patients in shock or with dehydration.

• Give only isotonic fluid (0.9% saline) by slow intravenous infusion to restore the circulating volume, but avoid circulatory overload, which may rapidly precipitate fatal pulmonary oedema.
— Monitor blood pressure, urine volume (every hour) and jugular venous pressure.
— Improve oxygenation by clearing the airway, increasing the concentration of inspired oxygen and supporting ventilation artificially, if necessary.</td></tr>
<tr><th>Coma</th><th>Treatment</th></tr>
<tr><td></td><td>• Maintain a clear airway
• Position the child in the lateral or semi-prone position to avoid aspiration of fluid (in case of trauma, stabilize neck first so that it does not move)
• If possible, insert a nasogastric tube and aspirate the stomach contents to minimize the risk of aspiration pneumonia, which is a potentially fatal complication that must be dealt with immediately.
• Assess level of consciousness using a paediatric scale (e.g. for children use the Blantyre coma scale, for adults use the Glasgow coma scale Annex VIII).
• Give intravenous (IV) glucose.
• Suspect raised intracranial pressure in patients with irregular respiration, abnormal posturing, worsening coma, unequal or dilated pupils, elevated blood pressure and falling heart rate, or papilloedema. In all such cases, nurse the patient in a supine posture with the head of the bed raised.
• Manage the underlying cause of loss of consciousness in children WITH fever:
— malaria, meningitis, sepsis.
• Manage the underlying cause of loss of consciousness in children WITHOUT fever:
— dehydration, anaemia, poisoning.</td></tr>
</table>

ANNEX VII

Assessment and treatment of danger signs

Signs	Check	If:	Treatment[a]
Convulsions	*Duration*	Lasts >5 minutes	Diazepam IV or per rectum
	Blood glucose	<2.2 mmol/l or test not possible	Give 50% dextrose IV
	Malaria slide or RDT	Positive or test not possible	Start antimalarial drugs
	Lumbar puncture (LP)[b]	CSF evidence of meningitis[c] or LP not possible	Start antibiotics for meningitis
Prostration	*Circulation* Capillary refill >3 seconds[d] Weak, fast pulse Cold hands	Any sign positive (indicates shock) and no evidence of severe malnutrition	Start rapid IV fluids Give oxygen
	Hydration Sunken eyes Lax skin turgor[e]	Any sign positive (indicates dehydration) and no evidence of severe malnutrition	Start rapid IV fluids or insert nasogastric (NG) tube and start oral rehydration solution
	Nutrition Visible severe wasting AND/OR Flaking skin and oedema of both feet	Any sign positive (indicates severe malnutrition)	Transfer to therapeutic feeding centre
	Blood glucose	<2.2 mmol/l or test not possible	Give 50% dextrose IV
	Lumbar puncture If none of the above signs is present, perform LP	CSF evidence of meningitis[c] or LP not possible	Start antibiotics for meningitis
	Malaria slide or RDT	Positive or test not possible	Start antimalarials

Coma	*Blood glucose*	<2.2 mmol/l or test not possible	Give dextrose 50% IV
	Perform LP	CSF evidence of meningitis[c] or LP not possible	Start antibiotics for meningitis
	Malaria slide or RDT	Positive[f] or test not possible	Start IV or IM antimalarials
	All comatose patients		Insert NG tube, aspirate stomach contents
Respiratory distress	*Palmar pallor* Check Hb	Hb <5g/dl	Give immediate blood transfusion
	Hydration Lethargy Sunken eyes Lax skin turgor[e]	Any sign positive and no evidence of severe malnutrition	Start rapid IV fluids
	Circulation Capillary refill >3 seconds Weak, fast pulse Cold hands	Any sign positive and no evidence of severe malnutrition	Insert IV and start rapid IV fluids
	Listen to chest	Chest crackles	Verify pulmonary oedema;[g] start antibiotics for pneumonia

[a] For treatment details see "Management of severe *P. falciparum* malaria" in chapter 6 of this document.

[b] Lumbar puncture should be performed if the patient is still unconscious more than 30 minutes after the end of a convulsion.

[c] CSF evidence of meningitis: CSF: blood glucose ratio <0.5, CSF protein >0.6 g/litre, bacteria seen on Gram stain. When measurement of glucose and protein is not possible, cloudy CSF should be taken as indicative of meningitis.

[d] Capillary refill: apply pressure for 3 seconds to whiten the fingernail; determine the capillary refill time from the moment of release to recovery of original nail colour.

[e] Skin turgor: pinch the skin of abdomen halfway between the umbilicus and the side for 1 second, then release and observe; if the skin takes >2 seconds to return, this indicates dehydration.

[f] In all cases of suspected severe malaria, parenteral antimalarial chemotherapy must be started immediately. If the cause of coma is in doubt, also test for (and treat) other locally prevalent causes of coma, e.g. bacterial, fungal or viral meningoencephalitis. If the malaria slide/RDT is negative it must be repeated, and the Giemsa-stained slide must be examined for indirect signs of parasitaemia (pigment in neutrophils).

[g] If pulmonary oedema is suspected, position the patient upright, give oxygen, stop IV fluids and give frusemide, 0.6 mg/kg IV. If there is no response, increase the dose progressively to a maximum of 200 mg.

ANNEX VIII

Assessment of level of consciousness

Children aged 5 years and below

The Blantyre coma scale[I] is a means of rapidly assessing and monitoring the level of consciousness in children. A score of 5 indicates full consciousness, while a score <3 indicates unrousable coma. Ideally, the assessment should be carried out 4-hourly in all unconscious children.

Response		Score
Motor	Localizes pain[a] Withdraws limb from pain[b] Nonspecific or absent response	2 1 0
Verbal	Appropriate cry Moan or inappropriate cry None	2 1 0
Eyes	Gaze oriented Gaze not oriented	1 0
Total		Maximum 5 (fully conscious) Minimum 0 (deep coma)

[a] Apply sternal pressure (the examiner should press firmly on the child's sternum using the knuckles of one hand). A child who is able to localize pain makes an attempt to remove the examiner's hand.
[b] Pressure applied with pencil to nail bed.

[I] Molyneux M et al. Clinical features and prognostic indicators in paediatric cerebral malaria: a study of 131 comatose Malawian children. *Quarterly Journal of Medicine*, 1989, 71(265): 441–459.

Children over 5 years of age and adults

Consciousness level should be assessed using the Glasgow coma scale:

Response		**Score**
Eyes open	Spontaneously	4
	To speech	3
	To pain	2
	Ncvcr	1
Best verbal response	Oriented	5
	Confused	4
	Inappropriate words	3
	Incomprehensible sounds	2
	None	1
Best motor response	Obeys commands	6
	Localizes pain[a]	5
	Withdraws to pain	4
	Flexion to pain	3
	Extension to pain	2
	None	1
Total		Minimum 3; Maximum 15[b]

[a] Apply sternal pressure (the examiner should press firmly on the patient's sternum using the knuckles of one hand). A patient who is able to localize pain makes an attempt to remove the examiner's hand.
[b] Unrousable coma is defined as a score of <11.

Ideally, the assessment should be carried out 4-hourly in all unconscious patients, to monitor the level of consciousness.

ANNEX IX

Rapid qualitative assessment of social, economic and cultural aspects of malaria

Introduction

This annex is intended as a guide only. The questions that will need to be asked to understand what community members know about, and how they respond to malaria, will depend on the local situation and on cultural norms. In addition, the approach used will depend on the subject group – key informants, householders, small groups, etc. In some settings there may be a need to focus just on treatment-seeking behaviours or on the use of preventive measures, while in other settings a more complete understanding of what malaria means to a community and what their priorities are may be needed. The approach should therefore be context specific. Translation of questionnaires (if necessary) can often be problematic. Pre-testing and validation of the questionnaire is also necessary.

The purpose of the rapid qualitative assessment is to provide information that will assist in making sound programmatic decisions related to malaria control.

Basic demographic data

Data on the age, gender, socio-economic status (such as education, literacy levels, etc) of the respondent should be collected. These data can be important explanatory variables, describing, for example, how knowledge varies by literacy. This, in turn, may influence the approach to public health awareness campaigns.

Community perceptions and knowledge of malaria

Community perceptions and knowledge of malaria can influence aspects of control. For example, in some areas of Africa, malaria is seen in the context of local beliefs of "magic" or "curses". This may heavily influence the impact public health messages have and the uptake of services such as treatment-seeking. Identifying local perceptions can therefore be useful in designing public health messages or strengthening informal treatment practices.

Some example questions to assess community perceptions and knowledge of malaria are given below, but these are only a guide and should be adapted to the local context.

1. What are the common illnesses in the community?
   Allow the person to list all illnesses, even if malaria or a malaria-like illness is not mentioned.
2. Which of these illnesses are considered to be the most important and why (prevalence, severity, mortality)?
3. Which of these illnesses affect children? Which affect adults?
4. What is the local name for "febrile illness"?
   Probe for terms for febrile illness.
5. Are there different types of febrile illness?
   a) If **yes**: What are the local names of these illnesses?
   b) Which of the febrile illnesses that you have described to me is considered to be the most dangerous?
   c) How do you know when a child/an adult has this type of illness?
   Identify which of the terms used for the illness most closely resembles the biomedical definition of malaria, then use that term for the questions below. In some cultures, fever with convulsions is considered to be a different type of malaria and one that is best suited for treatment by traditional or spiritual healers. If this is not mentioned, you could add a question such as, "Is there a different illness when the child/adult has fever and also convulsions (shaking)?"
6. Who is most at risk for [local term for malaria]?
7. What causes [local term for malaria]?
   a) Can malaria be spread?
   b) If **yes**: How is it spread?
   If mosquitoes are not mentioned, probe with questions like: "Can this illness be carried by mosquitoes?"

Treatment-seeking

These questions can be used to identify the channels through which treatment is sought. Treatment could, for example, be dominated by traditional/spiritual healers or by local pharmacies or drug shops, which can be included in programming to improve practices. The questions can also be used to identify barriers to access to appropriate treatment, such as cost.

8. How is [local term for malaria] treated?
   Probe to ensure that all treatments are mentioned, including use of self-medication.
   a) Can you please tell me the steps of treatment once you notice that somebody in your household/community is sick with [malaria]?
   b) If **use of health care services** is mentioned:
      - Ask about the last case in the household/community. For example, "I see you mentioned going to the [dispensary/clinic/hospital] to receive care. From the time that you knew that your child/the adult was sick, how long did you wait before going to the health care facility?"
      - What do you do for treatment during times when the [dispensary/clinic/health post] might be closed (such as weekends, nights, holidays)?
      - Is there an ambulance service to help you with a really sick person? *If the answer is* ***no****: probe to find out how they transfer a sick individual to a health care facility.*
   c) If **multiple treatments** are listed: Did you use some of these treatments at the same time (for example, see a traditional healer for herbal treatments while taking antimalarials)?
   d) Which of these treatments are considered to be the most effective?
   e) From whom do you normally receive advice about how to treat illness in a child/adult?
   f) **If not previously mentioned:**
      - Do you use the services of traditional healers or spiritualists for malaria?
      - Do traditional healers ever give you antimalarial drugs?
   g) Can you easily obtain drugs to treat [malaria] here in the camp/ settlement area?
      - If **yes:** Where do you get these drugs?
   h) Can you please tell me about any problems you might experience when seeking care from a health care facility in the camp/settlement area?
      Probe for things such as lack of referrals, no assistance to transport a sick patient, long waiting periods for care, political/cultural differences with health care staff, perceived poor communication between staff and patients or caregivers, lack of drugs or other supplies.

i) I would like to ask you about the cost of an episode of [malaria]. For the last case of malaria in the family, could you tell me how much money you spent on the following?
   - transport to or from a health facility
   - consultation with a doctor/nurse/traditional healer
   - diagnosis of the disease by a laboratory
   - drugs to treat the disease
   - other items (such as syringes, drips, etc)
   - food or other special items that were needed to help the patient
   - were there any other costs that you can remember?

Prevention

Perceptions on prevention of malaria are important because some areas will be used to conventional prevention methods (e.g. bednets) but some areas may never have been exposed to such interventions and people may be resistant to their use or inexperienced about how to use them effectively. Lack of previous exposure to malaria control interventions may result in misconceptions concerning their safety or effectiveness. Some locally used prevention methods may have no proven effect on malaria control and may actually be harmful (e.g. burning dung as a repellent). Collecting information on local perceptions of prevention can help in the understanding of: why certain public health messages do or do not work; the usage of interventions; and receptiveness to IRS.

9. Can [local term for malaria] be prevented?
   a) If **yes**: How can you prevent it?
      Probe for use of insecticides, bednets or other insecticide-treated materials, coils, repellents, etc.
   b) Of these methods, which ones do you personally use?
   c) Why do you use these particular methods?
   d) Which of the methods that you mentioned do you think are the most effective?

Insecticide-treated materials

10. Where you live, can you estimate for me how many households use LLINs?
    a) Would you say that where you live is similar to other parts of the settlement area?
    b) Did people bring the LLINs with them from their previous homes or did they receive them while they were in camp?

c) If **a net was mentioned** in response to question 9b): Where and when did you get your LLIN?
   - Do you know whether your net was treated with insecticide?
   - Have you ever had a LLIN while in the settlement that you needed to sell in order to buy something else, for example food or water?
   - If the **interview is done at home** and if the respondent states that they have a LLIN: Could you please show me your net?
     Look to see what type of net it is, for example, is it polyester, polyethylene or polypropylene? What is the condition of the net – is it dirty, is it very clean, does it have holes in it?

d) If a LLIN **was *not* mentioned** in response to question 9b): "I see that you did not mention using a LLIN. Why is this?"
   Probe for reasons of cost (cannot afford), lack of availability (such as distribution only to vulnerable populations), theft of nets, needed to sell the LLIN, dislike due to heat.

e) Why do people use LLINs?
   Probe for prevention of nuisance biting, prevention of malaria, prestige, privacy.

f) In your house, who sleeps under the LLIN?
   - Why does that person/those persons sleep under the LLIN?

g) Are there any times when households in this area do not use LLINs if they have them?
   Probe for example, if people sleep outside because it is too hot, do they sleep under the LLINs?

h) Do people always sleep indoors – or outside, or in fields? Does this change with the season?

i) Are LLINs supported by poles or are they suspended from ceilings?

j) What sizes and shapes of LLINs would best fit your sleeping arrangements?

Indoor residual spraying (IRS)

11. Have you ever had your house sprayed with insecticide to prevent mosquitoes/malaria?
    a) If **yes**: What did you think about that malaria control measure?
       Probe: Ask whether they like the control measure; did they fear the insecticide being in the house.
12. Do you think that this method works to control mosquitoes?
13. Was there anything that you did not like about IRS?

14. Once your house was sprayed with insecticide, did you replaster the walls? (How much time elapsed between the spraying of the walls and replastering?)

Experience with malaria control

Some communities will be highly sensitized to malaria control from past experiences, but some may be naive about the methods commonly used for control. This will influence the approach to implementation.

15. Did you have [local term for malaria] in your home area?
    a) If **yes**: What was your experience with malaria control?
       Probes: Ask whether they ever had indoor residual spraying done in their home; whether the health care facilities dispensed antimalarials; whether there were net distribution or insecticide re-treatment programmes taking place?
       - Was the community involved in trying to control malaria? If yes: What did they do?
16. Can you please give me some ideas for what you would like to see here to improve malaria control?

ANNEX X

Tube and cone assays for insecticide testing of adult mosquitoes

Figure X.1 **The WHO tube assay**

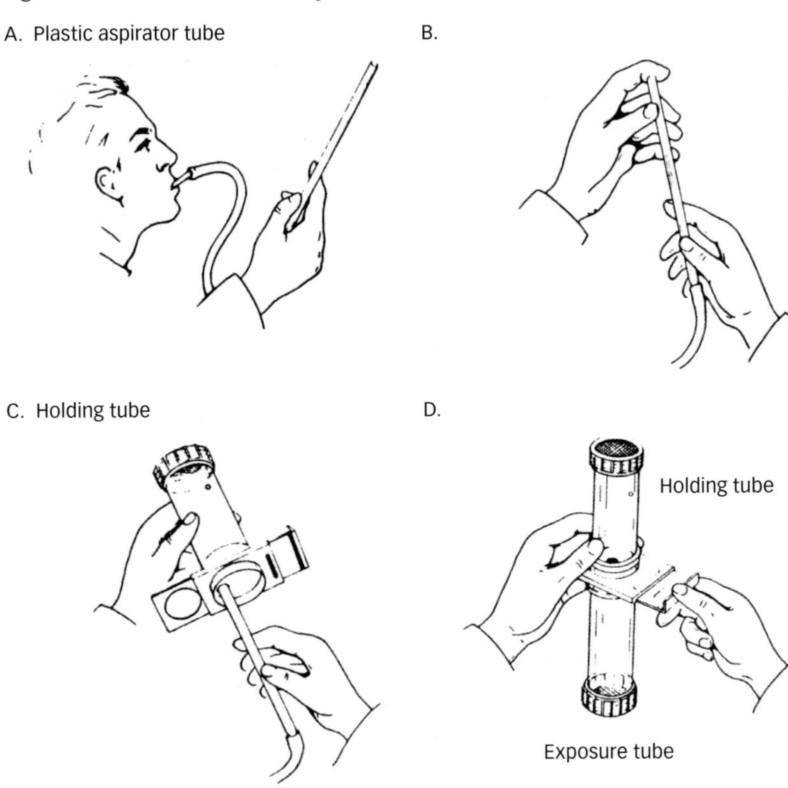

The WHO tube test kit consists of two plastic tubes (125 mm in length, 44 mm in diameter), with each tube fitted at one end with a 16-mesh screen. One tube (exposure tube) is marked with a red dot, the other (holding tube) with a green dot. The holding tube is screwed to a slide unit with a 20 mm hole into which an aspirator will fit for introducing mosquitoes into the holding tube. The exposure tube is then screwed to the other side of the slide unit. Sliding the partition in this unit opens an aperture between the tubes so that the mosquitoes can be gently blown into the exposure tube to start the treatment and then blown back to the holding tube after the timed exposure (generally one hour). The filter-papers are held in position against the walls of the tubes by four spring wire clips: two steel clips for attaching the plain paper to the walls of the holding tube and two copper clips for attaching the insecticidal paper inside the exposure tube. The purpose of this test is to determine if the local malaria vectors are susceptible to insecticides for treated nets and for IRS. See revised thresholds for recording resistance – http://www.who.int/malaria/vector_control/ivm/gpirm/en/index.html.

WHO cone assay

WHO cone assays should be conducted on 25 cm x 25 cm pieces cut from positions 2, 3, 4 and 5 of each sampled net (see Figure X.2), which should be adjacent to the places from which the netting for chemical assay was collected.

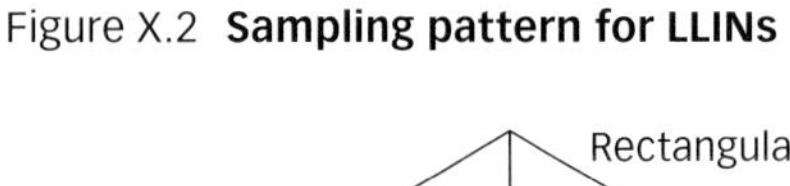

Figure X.2 **Sampling pattern for LLINs**

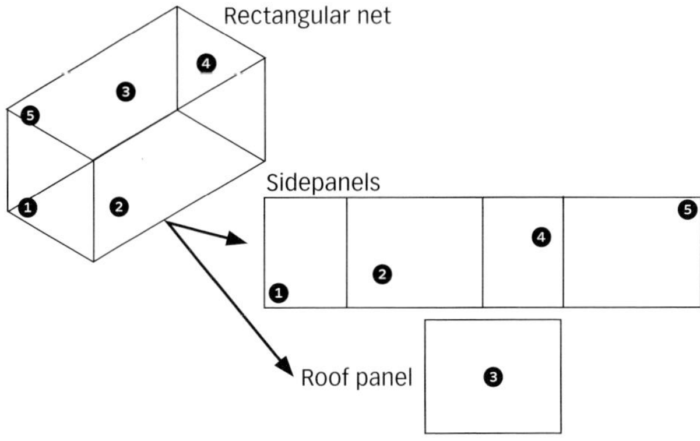

Position 1 should be excluded, as it may be exposed to excessive abrasion in routine use, as this portion of the net is frequently handled when it is being tucked under the bed or mattress. Two standard WHO cones are fixed with a plastic manifold onto each of the four netting pieces (Figure X.2) or in the case of IRS, fixed on sprayed wall surfaces. Between 5–20 susceptible, non-blood-fed, 2–5-day-old female Anopheles (species to be stated in the test report) are exposed for 3 min in each cone and then held for 24 h with access to sugar solution. Two replicates should be placed at each position. Knock-down is measured 60 min after exposure, and mortality is measured after 24 hours. A negative control, from an untreated net, should be included in each round of cone bioassay testing. If the mortality in the control is between 5% and 20%, the data should be adjusted with Abbott's formula.

If the mortality in the control is >20%, all the tests should be discarded for that day. Bioassays should be carried out at 27 ± 2 °C and 80 ± 10% relative humidity. The bioassay results for the netting pieces from each sampled LN should be pooled to determine if the net meets the WHO efficacy requirement, i.e. ≥80% mortality or ≥95% knock-down. If the net fails these criteria, a tunnel test should be conducted on one of the four net samples that caused mortality closest to the average mortality in the cone bioassay.

Finding out more

- WHO. (1998). *Test procedures for insecticide resistance monitoring in malaria vectors, bio-efficacy and persistence of insecticides on treated surfaces.* Geneva, World Health Organization, 1998 (WHO/CDS/CPC/MAL/98.12).
- WHO. (2011). *Guidelines for monitoring the durability of long-lasting insecticidal mosquito nets under operational conditions.* Geneva. World Health Organization.
- WHO (2012). *Global Plan for Insecticide Resistance Management in malaria vectors.* Geneva, Switzerland. World Health Organization
- WHO (2013). *Test procedures for insecticide resistance monitoring in mosquitoes.* Geneva. World Health Organization.

Figure X.3 **Example form for cone bioassay of LLINs collected in houses**

Name of person performing bioassays: ....

1. Date of test (dd/mm/yyyy):....../......./.................
2. LN code: ....../....../....../....../....../....../....../......
3. Temperature: ....../...... °C   Relative humidity: ....../....../...... %
4. Test mosquito species:
5. Age of mosquitoes: ...... days
6. Test start time (h/min): ....../...... End time (h/min): ....../......

Net position	Replicates[a]	No. of mosquitoes exposed[b]	No. knocked down after 1 h	No. dead after 24 h	No. alive after 24 h	% knocked down	% mortality
1[c]	1						
	2						
2	1						
	2						
3	1						
	2						
4	1						
	2						
5	1						
	2						
Control	1						
	2						

[a] Two cones on each net sample (replicates 1 and 2)
[b] Usually, five mosquitoes per cone; exposure time, 3 min
[c] Net position 1 should be tested only in baseline bioassay

ANNEX XI

Example: malaria outbreak form

Table 4.1 **Example of malaria outbreak form**

Place: .... Reported by: ....

From:....../......./......... (day/month/year) To: ....../......./......... (day/month/year)

Case definition "malaria" and "malaria death":

Date	**Day**	**New malaria cases**				**New malaria deaths**			
		<5 years	**≥5 Years**	**Pregnant women**	**Total**	**<5 years**	**≥5 Years**	**Pregnant women**	**Total**
	1								
	2								
	3								
	4								
	5								
	6								
	7								
Total									

Daily mortality rate (number of "malaria" deaths/10 000 population per day): ....

Weekly attack rate (number of "malaria" cases/1000 population per week): ....

Weekly case-fatality rate (number of "malaria" deaths/number of "malaria" cases x 100): ....

Geographical areas affected (e.g. villages, districts, camps where malaria patients are currently living):

....

....

ANNEX XI. EXAMPLE: MALARIA OUTBREAK FORM

Locality	**Estimated total pop. in locality**	**No. of malaria cases from locality** (specify day, week, time period)	**No. of malaria deaths from locality** (specify day, week, time period)

Comments: ....

....

....

Index